ENVIRONMENTAL HAZARDS
Human and Policy Dimensions

VOLUME 9 ISSUE 3 2010
SPECIAL ISSUE

AF905571

Special issue: Coastal hazards and vulnerability

Environmental Hazards 9(3) September 2010.

First published 2010 by Earthscan

2 Park Square, Milton Park, Abingdon, Oxon OX14 4RN
711 Third Avenue, New York, NY 10017, USA

Routledge is an imprint of the Taylor & Francis Group, an informa business

First issued in hardback 2016

ISSN: 1747-7891 (print), 1878-0059 (online)
ISBN: 978-1-84971-211-8 (hbk)
ISBN: 978-1-138-99143-9 (pbk)

SUBSCRIPTIONS

Subscription prices for Volume 9:

Institutions

Online only:	£247	$494	€323
Online & print:	£260	$520	€340 (airmail extra)

Personal

Online only:	£99	$199	€130
Print only:	£99	$199	€130 (airmail extra)

Orders can be placed online at www.earthscan.co.uk/journals/ehaz or sent to the journal's distributors, Portland Customer Services, using the contact details below.

Post: Portland Customer Services, Commerce Way, Colchester, CO2 8HP, UK
Fax: +44 (0)1206 799331
Tel: +44 (0)1206 796351
Email: sales@portland-services.com

Abstracting services which cover this title include Elsevier Scopus and GeoRef

Environmental Hazards is published quarterly. Periodicals Postage Paid at Rahway, NJ. US agent: Mercury International, 365 Blair Road, Avenel, NJ 07001. POSTMASTER: Address changes to ENVIRONMENTAL HAZARDS, 365 Blair Road, Avenel, NJ 07001.

GUEST EDITORIAL

RESEARCH

Routledge
Taylor & Francis Group
LONDON AND NEW YORK

earthscan
from Routledge

ENVIRONMENTAL HAZARDS
Human and Policy Dimensions

ENVIRONMENTAL HAZARDS: **Human and Policy Dimensions** (EHAZ) is an innovative, interdisciplinary and international research journal addressing the human and policy dimensions of hazards.

The journal addresses the full range of hazardous events from extreme geological, hydrological atmospheric and biological events such as earthquakes, floods, storms and epidemics to technological failures and malfunctions such as industrial explosions, fires and toxic material releases. EHAZ is the source of the new ideas in hazards and risk research.

With a genuinely international perspective, this journal highlights issues of human exposure, vulnerability, awareness, response and risk. The role of hazards in development, and issues of efficiency, social justice and sustainability are also explored in the journal.

Well known conventional hazards receive extensive coverage but submissions about new forms of hazard, emerging risk management institutions and restructuring of ideas about hazards – including their role in human affairs – are particularly welcome.

Reinvigorating the debate about how we define, understand and manage hazards, the journal is interdisciplinary in scope and open to contributions by specialists from a wide range of fields who are interested in the effects of hazards events on people, property and societies.

ENVIRONMENTAL HAZARDS
Human and Policy Dimensions

Coastal hazard, vulnerabilities and resilience

LORAINE MCFADDEN*

Flood Hazard Research Centre, Middlesex University, London, UK

1. Introduction

This volume seeks to explore environmental hazards in a coastal context. We know in this respect that populations are moving to the coast on a global scale, and development there is faster than in inland locations. The risk of hazard is growing, and this is occurring in both the developed and developing parts of the world, not least owing to the threat of global sea-level rise (Nicholls et al., 2008). The past decade has seen major disasters in coastal locations, such as Hurricane Katrina and the Indian Ocean Tsunami, and no doubt others will follow.

Therefore, there is good evidence that coastal communities are increasingly at risk, from flood and erosion, exacerbated by the kind of land subsidence that is affecting or has affected most megacities at the coast such as Bangkok and Shanghai, itself a result of the intensifying human occupancy of coastal regions. Increased vulnerability to the hazards that are experienced in coastal areas is the result.

But we see the concept of coastal vulnerability as based on human value judgements concerning risk to various elements of the natural and human environment from a variety of sources, not as an absolute (Green and McFadden, 2007). Therefore, the vulnerability of a coastal community is a function of the state we wish that community to be in, its relations with other communities, the relevant governance arrangements at the coast in question, and linkages or integration with the natural environment of the space it occupies.

Thus in developing this volume we have been minded to identify research that illuminates these aspects of coastal hazardousness in different situations and with different emphases. In so doing we have followed the key themes that *Environmental Hazards* pursues: human dimensions and policy implications. In this respect, we look, below, at some aspects of the six papers that they share in common, starting with the concepts of resilience and the different scales at which it needs to be recognized.

2. Concepts and scales

Lesley Ewing and her colleagues remind us that the dynamism of the coast is one of its enduring features, and that the ever-changing water levels and shifting shoreline conditions are part of people's fascination with the coast as well as the source of its hazardousness. Change at the coast is inherent, not to be unexpected but part of regular coastal processes, and the difference between a normal coastal event and a hazardous or extreme event is only a matter of degree and the rare occurrence of coinciding events.

Indeed, we need to remember that the coast has been an area of significant hazards throughout human history and that in modern times at least hazard responses have typically been focused on resistance – constructing stronger buildings, enhancing natural barriers or creating artificial barriers between the land and the sea. However, interest in resilience has grown as

■ *Email: L.McFadden@mdx.ac.uk

doi:10.3763/ehaz.2010.0055 ISSN: 1747-7891 (print), 1878-0059 (online) www.earthscan.co.uk/journals/ehaz earthscan

resistance alone has come to be seen as less effective than is desirable, or even unsustainable, in the longer term.

This paper also makes the key point that resilient communities tend to learn from past experience; each disaster provides valuable lessons to avoid development in high hazard areas or to redevelop near the coast in ways that will lessen consequences of the next event. The dilemma here is that those without this experience do not learn so readily, and so are trapped by their inexperience into pursuing or maintaining behaviours that exacerbate risk.

Lesley and her colleagues come from an engineering background and stress that the role of engineering technologies remains important, especially in monitoring risk trends using modern methods. Science is generally mature in these areas, but the continuous improvement in hazard modelling coupled with the rigorous use of field surveys can be used to identify policy options. They can also provide the foundation for improved design guidance for the development and infrastructure that needs to be located in coastal hazard areas.

All the papers in this volume stress the importance of spatial scale and its effect on the analysis of both vulnerability and resilience: viewed from a global perspective, local vulnerabilities may shrink in importance, even if they are acutely felt by the communities affected. These differences are pursued more systematically than has been common in the past by Suzanne McLaughlin and Andrew Cooper in developing vulnerability indices incorporating a diversity of indicators to provide different levels of spatial analysis of vulnerability. What this shows, naturally, is that some important local variations in vulnerability are masked by the necessary simplifications that map vulnerability at a national scale. Data availability is also a key issue here: for some variables more detailed information is available as the spatial resolution of an analysis increases, while others become obsolete as data are of insufficient resolution to differentiate real variability at more detailed scales.

This research highlights that while a common index architecture can be applied, the selection of variables must take account of the scale at which the hazard is to be assessed. The results show the value of a nested approach to index development and the scale-dependent nature of policy response. A national-level vulnerability analysis gives a broad overview of the relative vulnerability of the coast, to be used, for example, in the implementation of coastal planning restrictions or in the allocation of money to be spent locally on coastal defences. A regional scale index can provide the context for the development of setback lines or hazard zones to control coastal development. At the local level, detailed analysis can identify infrastructure in potentially highly vulnerable areas. They stress that it is therefore the responsibility of the coastal manager or policy maker to learn what is the most appropriate scale or scales at which to investigate the vulnerability that they observe, depending on whether the policy 'push' or hazard response imperative is a national-, regional- or local-level one.

3. Communication and networks

Two papers in this volume concentrate on the importance of communication, both in the immediate hazard situations and in pre-planning for risk reduction in the promotion of resilient communities. In doing this, Frank Thomalla and Rasmus Larsen reiterate the 'old chestnut' that the technological aspects of early warning system development have commonly been receiving considerably more attention than human aspects.

They regret the situation, as they see it, that even though the importance of addressing community linkages in warning systems is strongly emphasized in current guidance from international agencies, practitioners still face considerable challenges in applying these insights in their operational contexts. Their research conducted a multi-stakeholder participatory assessment to create an improved understanding of the challenges and enabling conditions for practitioners to implement such recommendations and guidance.

In this paper, the canvas is broad. Insights from Thailand, Sri Lanka and Indonesia are used to analyse selected elements of resilience based on the Coastal Community Resilience (CCR) framework (US-IOTWS, 2007). Their findings indicate, first, that developing participatory and people-centred early warning systems has not been sufficiently translated into action in the implementation of national policies and strategies for early warning.

They also conclude, second, that policy and guidance place significantly more emphasis on the procedural compared to the normative and cognitive dimensions of early warning systems and that practitioners engaged in early warning and disaster risk reduction operate in contexts shaped by multiple stakeholder agendas and face considerable challenges in negotiating diverse needs and priorities. Many issues are outstanding, in their view, not least that few arrangements exist that enable stakeholders 'to coordinate and reconcile agendas, negotiate joint targets, share knowledge and critically reflect on lessons learnt, and to improve the integration of early warning with other priorities such as livelihoods improvement, natural resource management, and community development'. Clearly, there is much more that needs to be done to make their coastal communities safer than they are now.

In my paper in this volume, I also take up the theme of communication but have focused my research on communication networks. Exploring ideas from the Institutional Analysis and Development Framework (Ostrom, 1990) within the South Australian coastal management context, I have identified a range of inter-agency interactions as potentially significant drivers for increasing the resilience of the coastal environment.

I work, as before (Green and McFadden, 2007), with the understanding that resilience is a function of a normative statement on the characteristics desirable in a functioning coastal system. One of my conclusions is that cross-scale interactions and informal relationships within and between users and managers are key interactions defining resilience outcomes within the current system in the case study that I investigated. Also a key conclusion from this exploratory analysis is that resilience to hazard could be substantially improved by a better understanding of, and facilitation between, informal and formal knowledge and processes of decision making. I also reiterate (see McFadden et al., 2009) that careful, well-constructed analysis of the physical, social and institutional systems at the coast is essential for policy progress.

4. Promoting resilience through economic development

Resilience to hazard is promoted through economic development: it is poor communities that cannot suffer the harm that hazards bring and cannot quickly recover from the economic downturns that result. Two very different case studies form the final papers in this volume, and each illustrates this phenomenon in different ways.

The first of this pair of papers, by Audun Sandberg, analyses coastal farming in Tanzania. In this respect, when reviewing coastal zone management and constructing its research base, we must not forget that most coastal areas are not developed as urban centres or as the focus for recreation but are valued and valid as locations for agriculture: providing the world with food from fertile flat floodplains and deltas. These areas are also vulnerable to hazard and, if affected, suffer the diminution of output that harm their farmers and reduce the globe's ability to feed its increasing population.

Audun Sandberg's paper, in this context, stresses that the transformation of risky floodplain systems into socially controlled environments was the result of an intricate interplay between ecology, demography, religion, social organization and the technology of the time. We need to see coastal agriculture, and its vulnerability, in this broad historical context, and Audun's case study research investigates one of the few natural deltas and floodplains left to study in the warm regions of the world: the Rufiji Coastal Delta and Floodplain.

In contrast to many other deltaic areas of the world, where engineering-based irrigation and drainage have been the culture of intervention designed to promote agricultural intensification, here a multi-crop agriculture has evolved based on rice, maize, cotton and peas. This arrangement, locally developed rather than imposed from without, has created a robust risk-minimizing system that has endured, helped by Arab, German and British colonization attempts to 'modernize' the system, resulting in new crops and varieties being incorporated in a way that made it even more robust.

As the research shows, this robustness – in the face of the risks involved in floodplain farming – comes from the flexibility of the system and its transparent feedback mechanisms, the latter facilitating learning. If the system is disturbed by a hazardous event, it is self-correcting within the same agricultural year and thus also has a high resilience: if a major flood drowns all the rice in April, 'large areas of moist and fertile soils can be planted with cotton, maize and cow peas'. But before we rush to commend this system without caveats, we need to remember that the research also shows that the economic return to labour will always be low in such a system, despite its apparent robustness. Audun concludes that, 'It takes a lot of extra work to constantly prepare for multiple futures. Low labour productivity is therefore the premium the Rufiji peasants have to pay for their complicated risk insurance system'. So here is no panacea for development elsewhere if we wish to promote higher incomes for those many impoverished agricultural communities at similar risks elsewhere in the world.

The last paper in this volume, by Deborah Peel and Greg Lloyd, also takes a case study approach, in this instance to critically discuss how economic regeneration in the coastal resort of Porthcawl, South Wales (UK), has integrated flood defences into part of a strategic spatial planning approach to the redesign of the town's seaside, promenade and harbour quarter. This integrated policy coupling approach is part of a plan to upgrade the town's waterfront and enable regeneration of the kind of run-down small seaside resort that is so common in Britain now that those seeking holiday locations prefer the sun of the Mediterranean rather than the bracing weather of the typical English summer.

The bipolar arrangement of two interlinked policies for hazard reduction and regeneration means that they have together contributed both to risk reduction and to promoting resilience in a way that individually they could not. The 'policy coupling' analysis research shows that this twin track process has sought to raise public awareness of the issues and risks associated with the town's coastal location and to integrate flood defence aspects within the strategic regeneration plan. This is designed to promote a greater sensitivity to climatic risks and potential flooding. The research reminds us of the inherent adaptive capacity of humankind, in this case of coastal communities, and identifies the need to accommodate in our coastal towns and cities a new civic awareness of potential environmental hazards in articulating the public interest and balancing the trade-offs involved in planning their futures.

5. Some conclusions

I have two main conclusions from editing this research, one concerning methodology and the other concerning learning.

Concerning methodology, much of the research reported in this volume necessarily and usefully investigated coastal case studies, because locally derived information is essential for understanding – and advocating – relevant policy options. Indeed we have no apologies here. These case studies, as I see them, have been conducted within a general theoretical framework that recognizes the complexity defining how combinations of variables affect the behaviour of the coastal system. They illustrate local or regional issues, but do not attempt to over simplify. Above all, they show how hazard and coastal management needs to be evidence based, and policy development needs to be grounded in reality rather than the abstractions of over generalized modelling and over simplistic master plans.

But what about learning? In my view, one major conclusion is that we have moved in hazard management at the coast beyond the state where risks are inadequately assessed owing to paucity of technique or lack of data (although the latter often remains a constraint). We no longer do not know what the problems are; we do not know how to implement the solutions. But what we do know is that social learning of communities and professionals is vital for vulnerability reduction and in promoting resilience, and that governance arrangements need to facilitate this learning process. In general, we are dealing with a human problem, not a technical issue. Social learning is, of course, not easy in situations where motivation is low because hazard events are sporadic and infrequent or where other imperatives dominate. Nevertheless, it remains clear that communities and their institutions need to remain vigilant and learn from their experience and that of others. Otherwise the hazards of the future will simply replicate those that we have seen in the past. If our research can assist this understanding and learning, then it will have been of some value.

References

Green, C. and McFadden, L., 2007. Coastal vulnerability as discourse about meanings and values. *Journal of Risk Research*, 10. 1027–1045.

McFadden, L., Tapsell, S. and Penning-Rowsell, E., 2009. Strategic coastal flood risk management in practice: actors' perspectives on the integration in flood risk management process in London and the Thames Estuary. *Journal of Ocean and Coastal Management*, 52(12). 636–645.

Nicholls, R. J., Wong, P. P., Burkett, V. R., Woodroffe, C. D. and Hay, J., 2008. Climate change and coastal vulnerability assessment: scenarios for integrated assessment. *Sustainability Science*, 3(1). 89–102.

Ostrom, E., 1990. *Governing the Commons: The Evolution of Institutions for Collective Action*. Cambridge University Press, Cambridge.

US-IOTWS, 2007. *How Resilient is Your Coastal Community? A Guide for Evaluating Coastal Community Resilience to Tsunamis and Other Coastal Hazards*. U.S. Indian Ocean Tsunami Warning System Program supported by the United States Agency for International Development and Partners, Bangkok, Thailand.

research paper

ENVIRONMENTAL HAZARDS
Human and Policy Dimensions

A review of coastal community vulnerabilities toward resilience benefits from disaster reduction measures[1]

LESLEY EWING[1,2,*], REINHARD E. FLICK[3,4] AND COSTAS E. SYNOLAKIS[1]

[1]Viterbi School of Engineering, University of Southern California, Los Angeles, CA, USA
[2]California Coastal Commission, 45 Fremont Street, Suite 2000, San Francisco, CA 94105, USA
[3]California Department of Boating and Waterways, Sacramento, CA, USA
[4]Scripps Institution of Oceanography, La Jolla, CA, USA

The coast has always been an area of significant hazards. In situations of community self-sufficiency, consequences of coastal hazards might be isolated to regions directly affected by the hazard. But, in the current global economy, fewer and fewer communities are isolated; damage to one location frequently has consequences around the globe and coastal community resilience can have broad-reaching benefits. Hazard responses for the built coastal environment have typically been resistance: constructing stronger buildings, enhancing natural barriers or creating artificial barriers. These approaches to hazard reduction through coastal engineering and shoreline defence efforts have been crucial to sustained coastal development. However, as coastal forces continue or magnify and resources become scarcer, resistance alone may be less effective or even unsustainable, and interest in resilience has grown. Resilience is a community's ability either to absorb destructive forces without loss of service or function, or to recover quickly from disasters. Community resilience encompasses multiple elements, ranging from governance to structural design, risk knowledge, prevention, warning systems and recovery. This paper focuses on hazards of coastal communities, and provides a review of some recent engineering efforts to improve the resilience elements of risk knowledge and disaster warnings for coastal disaster reduction.

Keywords: coastal disasters; community; hurricanes; resilience; tsunamis; vulnerability

1. Coastal hazards and coastal disasters

Ever-changing water levels and shifting shoreline conditions are part of people's fascination with the coast and sources of its hazards. Gently breaking waves can be hypnotic; large storm waves can be treacherous. Dynamic changes to the coast are part of regular coastal processes. The difference between a normal coastal event and a hazardous or extreme event is a matter of degree and the rare occurrence of coincident events. For example, the 1982/1983 El Niño storms were quite damaging to the western US coast; the worst damage occurred when distantly generated swell combined with local storm waves during a period of high tides (Flick, 1998). Yet, in 1988, one of the most intense eastern Pacific storms ever recorded deposited debris along the coast, did a lot of cosmetic damage, yet caused only minor structural damage primarily because landfall did not coincide with high surge or high tide (Armstrong and Flick, 1989; Flick and Badan-Dangon, 1989).

Coastal disasters develop when hazard events lead to a significant loss of life or property or reduction in the quality or quantity of resources that are valued by people. Cutter (2001) defines a disaster as a 'singular event that results in widespread losses to people, infrastructure, or the environment'. As defined by the International Society for Disaster Reduction, a disaster is a 'serious disruption of the functioning of a

■ *Corresponding author. *Email:* lewing@coastal.ca.gov

doi:10.3763/ehaz.2010.0050 ISSN: 1747-7891 (print), 1878-0059 (online) www.earthscan.co.uk/journals/ehaz earthscan

community or society causing widespread human, material, economic or environmental losses which exceed the ability of the affected community or society to cope using its own resources' (ISDR, 2004). Hazard events, even extreme events, are not considered disasters if there are no impacts to humans or the human-valued environment.

People have settled in coastal areas for millennia despite the occurrence of coastal storms, floods, high waves, hurricanes, tsunamis, shifting coastal land through erosion, accretion, subsidence and uplift, and other hazards. The numerous defence, trade and commerce benefits and resources available to coastal communities have outweighed the risks from coastal hazards; nevertheless, disasters are regular coastal occurrences.

Coastal disasters are triggered mostly by natural events. Development in vulnerable coastal lands can intensify consequences of hazard events and escalate damaging events to disasters. Human activities can also alter the severity of events; for example, increased emissions of greenhouse gas have been linked to increased sea levels; increased groundwater levels can reduce slope stability and increase bluff erosion; increased hardscape in watersheds and building in floodplains can intensify flood events; diversion of surface water can change flood patterns; flood control structures can alter sedimentation patterns and reduce coastal sediment supplies; and removal of coastal dunes, mangroves and offshore reefs can increase exposure of coastal areas to waves and storms (Griggs and Savoy, 1985; Griggs et al., 2005; Synolakis and Kong, 2006).

The 21st century has already experienced several coastal mega-disasters – the Indian Ocean tsunami, Hurricane Katrina and Cyclone Nargis. Each resulted in thousands of fatalities, social impacts and enormous financial losses from damage or destruction of buildings and infrastructure, disruptions to businesses and lost livelihood. While every disaster is unique, there are many similarities between disasters and each highlights the need for improved community resilience to address future hazard events.

The 2004 Indian Ocean tsunami was responsible for over 182,000 confirmed fatalities, 130,000 people missing, 1.69 million people displaced and US$14 billion in economic losses (Guy Carpenter Inc., 2005). More than a dozen countries around and near the Indian Ocean experienced directly the devastation of this coastal event. Many more countries experienced indirect losses – fatalities or injuries to people who were working in or visiting the tsunami areas, business losses and supply chain interruptions. Effects also extended to countries that provided disaster relief and aid or that experienced reduced aid benefits as resources were diverted to the tsunami-impacted areas. The magnitude of the tsunami and the scale and range of the impacts made this coastal event a disaster of global significance.

Hurricane Katrina was more geographically focused than the Indian Ocean tsunami; however, as it travelled through the Caribbean, Gulf of Mexico and inland across New Orleans and central USA, this Category 4 hurricane caused an estimated 1,500 deaths and US$81 billion in damages (Beven et al., 2008). Towns and cities throughout the Gulf region were destroyed. Following Hurricane Katrina many of the ports recovered quickly from hurricane damages, but found that their employees were not able to return quickly to work because they had nowhere to live (Curtis, 2007). The same was true for many other commercial and industrial activities. The regional shipping and commercial impacts from Hurricane Katrina lasted far longer than from the facility closures, due to the extensive community losses throughout the region. Since the shipping and commercial activities in the Gulf area are part of a regional, national and, in some cases, global economic network, the impacts from the loss or slow-down of shipping and commerce extended far beyond the main impact areas. The loss of local housing had consequences extending far beyond the individuals who experienced the direct losses.

Global connectivity is expanding the significance of coastal disasters from a broad range of hazards. Cyclone Nargis was responsible for more than 138,000 fatalities, 2.4 million people impacted, and US$4 billion in asset damages and lost income (Tripartite Core Group et al., 2008; Fritz et al., 2009). Myanmar's policy of economic isolation perhaps minimized impacts on other regions; however, both the Indian Ocean tsunami and Hurricane Katrina caused social and business disruptions that rippled throughout the world.

Coastal disasters can occur from a broad range of events and not just the extreme. Hurricane Ike was a spatially broad, Category 2 hurricane. There were 82 Ike-related fatalities in the USA and another 83 in the Caribbean. US communities on the upper Texas coast were inundated for days, businesses were closed, energy systems were shut off, and the economic consequences of these impacts rank Hurricane Ike as the third most costly hurricane in the USAs, following Hurricane Katrina and Hurricane Andrew (FEMA, 2008b). It seems only a matter of time before this ignominious ranking is surpassed. Current climate research anticipates that many coastal hazards will worsen in the future due to an increase in global temperature, rising sea level, and possibly increases in the intensity and frequency of storms and hurricanes.

2. Community vulnerability and response to coastal disasters

Up until late in the 19th century, communities viewed coastal disasters as unexpected, randomly destructive events or 'acts of god'. Most coastal regions now have well-recognized risks from coastal hazards, and few coastal disasters can be considered either unexpected or unprecedented. All three of the 21st-century coastal disasters were anticipated in theory, even if the exact time and place are unknown. Tsunamis have occurred prior to 2004 in the Indian Ocean. Seismologists had not identified the segment that ruptured in 2004, but had recognized the tsunamigenic potential of the Sunda Subduction Zone that triggered the Indian Ocean tsunami. On the island of Simeulue, near Banda Aceh, lessons from past tsunamis had been learned by new generations through ancestral teaching; thus people on the island recognized the harbinger signs of a possible tsunami and knew to evacuate to high ground. There were only nine fatalities out of a community of 83,000, despite inundation levels exceeding 10 m at the northern part of the island and up to 4 m along the southern part (Jaffe et al., 2006). Engineers with the New Orleans District of the US Army Corps of Engineers and local scientists discussed the vulnerabilities of the New Orleans levee system years prior to Hurricane Katrina. Just two years before Cyclone Nargis, another Very Severe Cyclonic Storm[2] made landfall in Burma, demonstrating the area's vulnerability to cyclones (Tripartite Core Group et al., 2008).

Many coastal events have some general trend. Flooding occurs with sufficient regularity that many areas are now mapped for flood probability (e.g. the Flood Insurance Rate Maps prepared for communities throughout the USA by the Federal Emergency Management Administration Agency). And, while one hundred year events do not occur precisely at 100-year intervals, the event probability, if based upon a sufficiently lengthy record and if the current situation is similar to the antecedent conditions, provides information on the range of likely impacts from any one event.

The western Atlantic and Gulf of Mexico are areas with frequent hurricane activity. One hundred and eight years before Hurricane Ike, the city of Galveston had been totally destroyed by a hurricane. From 1851 to 2008, dozens of Category 1 and 2 hurricanes and nine Category 3, 4 or 5 hurricanes passed within 100 miles of Galveston, underscoring the vulnerability of the Galveston area, and most of the Gulf Coast to hurricanes. Almost every 'unexpected' coastal disaster has a precursor or pattern to anticipate the event; however, oftentimes precursors or trends are overlooked, thought to be outliers, or non-indicative of the local coastal environment. The only unexpected aspects of most coastal disasters are the timing, magnitude and intensity.

Certain coastal events such as hurricanes, tsunamis, storm waves, El Niños and erosion are now routinely considered hazardous due to their potential to become a disaster. Awareness and understanding of coastal processes that precipitate hazard events are important for vulnerability assessments as well as for risk reduction. Communities have little motivation to consider hazard mitigation or resilience if there is no acceptance that a hazard event is possible or no understanding of the associated risks and vulnerabilities, and hence are unlikely to support efforts to address them. Hazard awareness and understanding are fundamental to risk and vulnerability assessments and efforts toward disaster risk reduction.

With the recognition that coastal disasters are recurring dangers comes the awareness of and some motivation or opportunity to reduce community vulnerability to future events. Concern about risks or vulnerabilities to disasters can be based on an awareness of local historic events or disasters from events that occur in distant, but similar areas. For example, the Indian Ocean tsunami, generated by the Sunda Subduction Zone, increased concerns for tsunamis from the similar Cascadia Subduction Zone. The worldwide extent of losses also led to studies of the potential consequences of Atlantic Ocean tsunamis. Specifically, this brought to attention an investigation of the structural evolution of the Cumbre Vieja volcano on La Palma, Canary Islands, and the potential for an Atlantic-basin-wide tsunami (Ward and Day, 2001). Communities with no recent history of tsunamis began to incorporate tsunami preparedness into emergency planning efforts.

The initial community responses to dangers from coastal forces have tended to parallel or mimic responses to other forces – to resist the forces through community fortifications. As noted by Godschalk (2003), 'traditional hazard mitigation programs have focused on making physical systems resistant to disaster forces'. Such an approach usually amounts to 'building for the last disaster' because estimates of impact metrics from the most recent disaster typically become the design standard for the physical resistance that will be part of the community's future protection. Structural resistance alone can be effective as long as the forces are less than the structural design limits, the design criteria have been enforced, and the structures are maintained to sustain design loads.

One of the recurring lessons from recent coastal disasters has been that resistance alone will not protect communities from disasters and that communities need to increase their resilience to hazard events. Current concerns for climate change have accentuated the need for flexible and adaptive coastal protection measures because rising sea level will heighten the effects of future hazard events (IPCC, 2001, 2007; Cayan et al., 2008).

Following the 2004 tsunami, the US Indian Ocean Tsunami Warning System Program outlined the range of community resilience outcomes for governance, resource management, land use, risk knowledge, warning, evacuation and disaster recovery, and recommended general resilience benchmarks for policy and planning capacity, physical and natural system capacity, social and cultural capacity, and technical and financial capacity (US Indian Ocean Tsunami Warning System Program, 2007).

The ASCE Hurricane Katrina External Review Panel (2007) made recommendations for a better understanding and communication of risks, taking a dynamic risk-based approach to hurricane and surge protection, upgrading the main levees and floodwalls, designing protection systems both for redundancy and in order to maintain functionality when overtopped (i.e. avoiding catastrophic failure), developing clear lines of responsibility, improving interagency coordination and updating engineering design procedures. These recommendations did not use the term 'resilience'. However, the overall effect is to identify and promote elements for making New Orleans a coastal resilient community.

3. Community resilience to coastal disasters

Resilience has been used for many years to indicate a general characteristic of something that can recover readily from stress. In materials science, resilience is a material property 'to absorb energy when it is deformed elastically, and to return it when unloaded' like a spring (ASM International, 2002), and the resilience modulus can be quantified from yield stress, strain and the modulus of elasticity. The general idea of resilience as a system property emerged in the 1970s, through work by C. S. Holling and a group of resource scientists and ecosystem managers who established the Resilience Alliance. The first major publication of ecological resilience, 'Resilience and stability of ecological systems', introduced the concept to ecosystem management (Holling, 1973). In this work, Holling (1973, p. 21) compares management for stability with resilience, where a

> management approach based on resilience, on the other hand, would emphasize the need to keep options open, the need to view events in a regional rather than a local context, and the need to emphasize heterogeneity. Flowing from this would be, ... not the assumption that future events are expected, but that they are unexpected. The resilience framework can accommodate this shift of perspective, for it does not require a precise capacity to predict the future, but only a qualitative capacity to devise systems that can absorb and accommodate future events in whatever unexpected form they may take.

In this early differentiation of management approaches, the stability approach would focus on resistance and the resilience approach would focus on the flexibility and adaptability of the system.

Several decades after Holling's (1973) work, Walker and Salt (2006, p. 1) defined ecosystem resilience as 'the ability of a system to absorb disturbance and still retain its basic function and structure'. This later idea of resilience is in line with the computer networking concept of resilience, which is synonymous with system fault tolerance and defines resilience as 'the ability of a computer system to continue to operate correctly even though one or more of its components are malfunctioning..... (t)o provide and maintain an acceptable level of service in the face of faults and challenges to normal operations' (NTIA and Federal Telecommunication Standards Committee, 1996). This definition also introduces the idea that the functions of a system may be the critical aspects of the system, rather than its individual components. Thus, for a community, the resilience of mobility may be more important than the stability of an individual road segment.

System function is an important aspect of resilience as is the ability to accommodate uncertainty. This aspect of flexibility to changing conditions has made resilience popular for several different disciplines, from its start in ecological management to the studies of social systems, engineering and disaster planning, and maybe is partially responsible for the focus on resilience in recovery planning after the Indian Ocean tsunami and Hurricane Katrina (ASCE Hurricane Katrina External Review Panel, 2007; US Indian Ocean Tsunami Warning System Program, 2007).

The resilience of complex social systems, such as cities or metropolitan areas, has the same characteristics as ecosystem resilience – the ability to withstand and recover quickly from shocks. Coastal events such as Katrina have highlighted the problems of using resistance as the only disaster risk management effort. Only recently has resilience been a strong focus of coastal managers, in response to the frequency and growing costs of coastal disasters. Resilient communities 'combine seemingly opposite characteristics, including redundancy and efficiency, diversity and interdependence, strength and flexibility, autonomy and collaboration, and planning and adaptability' (Godschalk, 2003, p. 141). The benefit of resilience is coping with uncertainty, as noted for ecosystems, and the expectation is that resilient cities will fare better after a disaster than cities that are less flexible. There is certainty that coastal events will continue to impact coastal communities. However, event specifics remain uncertain; 'resilience is not tied to a specific pattern of urban form of development .... (allowing) response to unique conditions ... (and) creative thinking' (Godschalk, 2003, p. 138).

Resilience has been closely linked with both sustainability and adaptation, especially for coastal communities, yet none of these terms have been precisely defined. Sustainability and adaptability can be qualities of resilience communities. Sustainability is characterized by the ability to meet current needs without jeopardizing the ability of future generations to do the same; it covers the concept both of need and of possible limitations by society, technology or the environment to meet these needs (ISDR, 2005). And, as resource and financial limitations of resistance-only approaches have become more apparent, sustainability has become an aspect of any disaster risk management effort.

Adaptation has been defined as adjustments in natural or human systems to reduce harm or exploit beneficial opportunities (USAID, 2009). Adaptation had become a focus for addressing climate change, since it can be in response to actual or expected changes and impacts (USAID, 2009), since neither the amount of climate change and sea-level rise nor the time when those changes will occur are known. Adaptive measures for coastal regions often address physical actions that can generally be grouped into retreat, accommodation or protection. On a global scale, analysis of adaptation considered protection for all developed systems and some restoration or enhancement of beaches and wetlands with economic value (Parry et al., 2009). Some studies have expanded adaptation to governance, and social and economic areas, looking at disaster risk reduction planning and diversified livelihood within the context of adaptation (USAID, 2009).

Over the past two centuries, most destroyed cities have been rebuilt (Vale and Campanella, 2005), which may or may not indicate resilience. History is replete with non-resilient cities – cities like Atlantis that vanished completely when an enormous surge of water washed it away,[3] or like Ephesus that attempted to adapt to changing conditions but ultimately could not be sustained. Ephesus was first established in about 1200 BC on the banks of the Cayster river, close to the Aegean. River silt built a delta, shoaling the original harbour. In 281 BC, a new Ephesus was built 2.5 km southwest of the original location to adapt to the expanding delta. The city was damaged by several earthquakes, the delta continued to expand into the Aegean, the harbour became less and less sustainable, and the city lost its position as a centre of trade and commerce. The abandoned city and harbour are now 5 km inland. Knossos, the capital of Minoan Crete, was eventually abandoned following the 1628 BC eruption of the Thera volcano.

Ecosystem resilience recognizes that communities are dynamic systems with some direction or trend. They are in some stage of rebuilding, demolition, recovery or decline. Walker and Salt (2006) observed that some disturbances cause systems to collapse into qualitatively different states that, while resilient, may not be the desired system state. Anoxic zones in the ocean are examples of this type of resilience. Mayor Riley of Charleston, South Carolina, at the 2002 Solutions to Coastal Disasters Conference, offered his observations that disasters push communities where they are already headed. Penning-Rowsell et al. (2006) have observed that post-crisis policy change is often not new policy, but rather the acceleration of policies that had started prior to the crisis, and that there are 'signals' of the likely changes prior to the crisis or disaster. Both trends in development and trends in policy provide clues about disaster response. A community that is in a period of revitalization will persist after the disaster or use it as a catalyst for improvement. But, if it is in decline and services are not being maintained or updated, then the disaster will likely accelerate the decline.

This trend of a community is part of resilience, where a resilient community will develop or continue on a trend of sustainability and improved function. Return of a community to a non-sustainable condition or to the same level of vulnerability that existed prior to the disaster is not indicative of resilience. As a working definition, based on a combination of several definitions, resilient communities can:

- adapt to changing conditions without losing function (Godschalk, 2003)

- recover from random events in unexpected forms (Walker and Salt, 2006)
- rely upon local and regional resources for much of the recovery (Holling, 1973) and
- use learned experience to reduce vulnerability and future risk (Resilience Alliance and Mayor Riley).

Thus, in the context of coastal disasters to the built environment, a resilient community is represented by the following characteristics:

- survival of people living in the community
- ability of people to remain in the community during the event or return quickly
- infrastructure remaining or becoming functional soon after the event
- maintenance or enhancement of community amenity values such as ecosystems and recreational space.

For coastal communities, resilience will be derived from some combination of good engineering, good land-use planning, redundancy of critical systems, enhancement of natural systems that mute or buffer coastal forces, and some elements to resist coastal forces.

4. Coastal engineering and community resilience

Several international strategies have been established around community resilience. The 1994 Yokohama Strategy for a Safer World and the more recent Hyogo Framework have brought international attention to disaster resilience, recommending that 'disaster resilience lies in the knowledge of the hazards and the physical, social, economic and environmental vulnerabilities to disasters that most societies face, and ways in which hazards and vulnerabilities are changing in the short and long term followed by action taken on the basis of that knowledge' (ISDR, 2005). Coastal research continues to improve both the understanding of processes and the ways of reducing hazards, risk, disasters and vulnerability and is assisting in disaster resilience in several areas.

Two research areas that have been important for advancing the understanding and awareness of coastal hazards and resilience are analytic modelling of coastal processes and field investigations of extreme events. Modelling improves risk knowledge, assists early warning systems and, for some hazards, approaches the point where model output can be used to determine areas most likely to experience major forces once an event is detected. Field investigations of extreme events have become regular occurrences and good opportunities to improve both risk knowledge and future resilience (Synolakis et al., 2005).

Numerical process modelling improves the ability to anticipate consequences of coastal events and provides a better understanding of site-specific hazard potentials. Modelling can provide an understanding of complex systems and physical phenomena that might otherwise require years of physical scale model testing and field observations. Analytical models can also help in understanding the mechanics of certain coastal processes that are not easily or properly represented by physical models. Some processes present scaling problems for physical models that require compromising the dynamic similarity of one or more parameters and allow only qualitative analysis of conditions, with limited quantitative utility.

Once numerical models are verified and tested, they can be calibrated for local conditions and used to anticipate possible coastal events. Two such models, MOST (Method of Splitting Tsunami; Titov and Synolakis, 1998) and SLOSH (Sea, Lake and Overland Surges from Hurricanes; Jelesnianski et al., 1992), have been used extensively to model extreme event tsunamis and hurricane surges. Both models provide forecasting tools of potentially disastrous coastal phenomena. These models can be used to identify the consequences of extreme events under current conditions and to model runup, inundation and surge with accelerating sea-level rise.

In addition to identification of areas vulnerable to tsunami runup or hurricane surge, models can develop hazard event libraries. Each model run requires considerable computational time.

By storing model runs as synthetic events, when actual events occur, modellers can retrieve similar synthetic events, scale the potential runup or surge with the appropriate input factors, and develop a rapid prediction of areas that may be at risk from the oncoming event (Bernard et al., 2006; USAID, 2009).

Model predictions rarely replicate real-world events exactly. Their results are limited by which physical parameters drive the model and how well the driving parameters are understood. After 10 m-high tsunami waves struck Sissano Lagoon, Papua New Guinea, model analysis of the earthquake source could not replicate waves of the orientation or height that actually occurred. Field investigations and bathymetric surveys of the surrounding area identified a recent underwater slope failure. Additional modelling, using a landslide source, closely reproduced observed waves and supported the landslide as the tsunamigenic source (Synolakis et al., 2002).

Field investigations are needed to examine the consequences of extreme coastal events, provide data that can be used to re-verify or recalibrate models, and provide new appreciation of generating mechanisms that necessitate new model approaches. Also, most field investigations examine the on-land component of a coastal event, where the coastal forces are transferred to terrestrial or constructed systems. This transition zone is the most difficult to include in analytical or physical models, except for those times when nature uses the coast as a full-scale laboratory. 'The evaluation of terminal effects [e.g. the effects at the end of the wave's journey across the open ocean when it arrived at the coast] of natural hazards remains one of the holy grails of geophysical research' (Synolakis and Bernard, 2006, p. 2231).

Examination of recent disaster areas has provided benchmarked data for model use. Synolakis and Kong (2006) prepared a summary of field investigations from 16 different countries affected by the Indian Ocean tsunami, and also provided an array of observations that will contribute to future improved resilience. Dalrymple and Kriebel (2005) evaluated the performance of coastal structures in Thailand following the tsunami, and found that many of the reinforced-concrete, well-engineered buildings fared well, but scour was a significant problem for buildings with shallow foundations. Building elevation, a pass-through lower floor and orientation with the short building dimension to the water were other features that were important for building survivability. Well-designed seawalls provided some protection to inland property; 'low seawall deflected much of the momentum of the waves skyward, reducing the forces on the landward structures' (Dalrymple and Kriebel, 2005, p. 9). On Phi Phi, a poorly constructed, sandbag core seawall failed in many places due to scour, and little, if any, protective value was apparent.

Seawalls were not the only features that protected inland areas; natural coastal features were also important. Low-elevation sand dunes on Karon Beach, Phuket, are attributed with dissipating wave forces so that inland development experienced high water levels without damage from high velocity, whereas a hotel that graded fronting dunes to improve ocean views was completely destroyed (Synolakis and Bernard, 2006). Liu et al. (2005) observed in Sri Lanka that mining of coral could have contributed to inland penetration of waves, which caused over 1,000 fatalities when the Sumudra Devi passenger train was derailed and overturned.

A recent field investigation of the Galveston area following Hurricane Ike made some observations for hurricane surge that compare with those of Dalrymple and Kriebel (2005) for tsunami damage. Buildings and shore protection elevation were critical to the survival of developed areas; scour was a major factor that destabilized both buildings and shore protection, and buildings with poorly embedded foundations were at risk from both wave impacts and scour (Ewing et al., 2009).

There have been many positive outcomes from the field observations of these events in addition to the field data that will be used for model refinement. Some observations from the Indian Ocean tsunami provide guidance on types of engineering structures that might survive a future tsunami event. Putting aside questions about rebuilding in inundation areas, observations of building survival have served as models for consideration of vertical evacuation options for already built communities. Design guidelines for vertical evacuation were developed and a companion report for planners is being planned (FEMA, 2008a). As a result of field observations from Hurricane Ike, FEMA revised directions for scour, pile design and embedment depths. No one event will provide all the solutions to coastal disasters and community resilience, but each event can provide incremental improvements.

5. Conclusions

The expected growth of coastal population and economy combined with the continued occurrence of hazardous coastal events means that people and resources will be exposed to major hazard events in the future; human development patterns that turn these events into disasters will also persist. The trend of a community can be an important indicator of community resilience – a community that is working to improve sustainable functions and capacity is better equipped to recover quickly from a disaster than one that is in decline prior to disaster. Resilient communities also tend to learn from past experience to avoid experiencing damages from the same vulnerabilities time after time.

Recent coastal research efforts have sought to 'learn from past disasters' and improve community resilience. Some of these efforts have included improving risk knowledge from post-disaster field investigations and identifying modelling efforts, warning opportunities and coastal protection options that will allow communities to improve resilience to future hazard events. This paper provides a summary of some recent coastal science efforts that, in combination with the other community governance, policy, planning and social elements, can improve community resilience.

Each disaster provides valuable lessons to avoid development in high hazard areas, or to redevelop near the coast in ways that will lessen the consequences of the next event. These can include building structures to avoid, or be more resistant to, anticipated storm forces; relocating structures and infrastructure out of vulnerable areas; and providing better monitoring and warning to allow people to evacuate to areas of greater safety; providing more education so that people in high-risk areas understand the risks and available risk reduction options.

Real-time monitoring of oceanic and atmospheric conditions allows communities to be aware of impending disasters and enables effective evacuation programmes to get people out of harm's way. Modelling and field surveys identify developed areas that should be evacuated or undeveloped areas that could be left as open space to avoid intensification of development in high hazard areas. Modelling and field investigations also provide the foundation for improved design guidance for the development and infrastructure that needs to be located in hazard areas. These combine to provide greater survival during a disaster, the ability of people to stay in or return quickly to the community, and infrastructure that can remain functional or return more quickly to functionality, helping to make the community more resilient.

Acknowledgements

The authors would like to thank Loraine McFadden for asking us to prepare this paper and for providing constructive comments and direction during the review process. We also thank an unknown reviewer.

Notes

1. This paper has not been reviewed by the California Coastal Commission.
2. The classification of Very Severe Cyclonic Storm is equivalent to a Category 4 storm on the Saffir–Simpson scale.
3. Here we offer our interpretation of Plato's intention in Critias, where it is noted that the submergence of Atlantis occurred in a single day following

earthquake and flood, as one of the earliest records of community that was not resilient to coastal disaster.

References

Armstrong, G. and Flick, R. E., 1989. Storm damage assessment for the January 1988 storm along the Southern California shoreline. *Shore & Beach*, 57(4). 18–23.

ASCE Hurricane Katrina External Review Panel, 2007. *The New Orleans Hurricane Protection System: What Went Wrong and Why*. American Society of Civil Engineers, Reston.

ASM International, 2002. *Atlas of Stress-Strain Curves* (2nd edn). ASM International, Materials Park, OH.

Bernard, E. N., Mofjeld, H. O., Titov, V. V., Synolakis, C. E. and Gonzales, F. I., 2006. Tsunami: scientific frontiers, mitigation, forecasting and policy implications. *Philosophical Transactions of the Royal Society*, 364. 1989–2007.

Beven, J. L., Avila, L. A., Blake, E. S., Brown, D. P., Franklin, J. L., Knabb, R. D., Pasch, R. J., Rhome, J. R. and Stewart, S. R., 2008. *Annual Summary: Atlantic Hurricane Season of 2005*. Tropical Prediction Center (ed.), National Weather Service, Miami, 1109–1173.

Cayan, D. R., Bromirski, P. D., Hayhoe, K., Tyree, M., Dettinger, M. D. and Flick, R., 2008. Climate change projections of sea level extremes along the California coast. *Climatic Change*, 87(Supplement 1). S57–S73.

Curtis, S. A., 2007. *Hurricane Katrina Damage Assessment: Louisiana, Alabama, and Mississippi Ports and Coasts*. American Society of Civil Engineers, Reston.

Cutter, S. L., 2001. *American Hazardscapes: The Regionalization of Hazards and Disasters*. Joseph Henry Press, Washington, DC.

Dalrymple, R. A. and Kriebel, D. L., 2005. Lessons in engineering from the Tsunami in Thailand. *Bridge*, 35(2). 4–13.

Ewing, L., Stauble, D., Work, P. A., Edge, B. L., Rogers, S. M., Loeffler, M. U., Kaihatu, J. M., Overton, M., Waters, J. P., Suzuki, K., Dean, R. G., Garrett, M. H., Wiggins, E. and Gregory, G. H., 2009. Field investigation of Hurricane Ike impacts to the upper Texas coast. *Shore & Beach*, 77(2). 9–23.

FEMA, 2008a. *Guidelines for Design of Structures for Vertical Evacuation from Tsunamis*. FEMA, Washington, DC.

FEMA, 2008b. *Hurricane Ike Impact Report*. Department of Homeland Security, Washington, DC, 57.

Flick, R. E., 1998. Comparison of California tides, storm surges, and mean sea level during the El Niño winters of 1982–83 and 1997–98. *Shore & Beach*, 66(3). 7–11.

Flick, R. E. and Badan-Dangon, A., 1989. Coastal sea level during the January 1988 storm off the Californias. *Shore & Beach*, 57(4). 28–31.

Fritz, H. M., Blount, C. D., Thwin, S., Kyaw Thu, M. and Chan, N., 2009. Cyclone Nargis storm surge in Myanmar. *Nature Geoscience*, 2(July). 448–449.

Godschalk, D. R., 2003. Urban hazard mitigation: creating resilient cities. *Natural Hazards Review*, 4(3). 136–143.

Griggs, G. and Savoy, L., 1985. *Living with the California Coast*. Duke University Press, Durham, NC.

Griggs, G., Patch, K. and Savoy, L., 2005. *Living with the Changing California Coast*. University of California Press, Berkeley.

Guy Carpenter Inc., 2005. *Tsunamis: Indian Ocean Event and Investigation into Potential Global Risks*.

Holling, C. S., 1973. Resilience and stability of ecological systems. *Annual Review of Ecology and Systematics*, 4. 1–23.

IPCC, 2001. *Climate Change 2001: Working Group I: The Scientific Basis*. Intergovernmental Panel on Climate Change.

IPCC, 2007. *Climate Change 2007, Synthesis Report*. Intergovernmental Panel on Climate Change.

ISDR, 2004. *Terminology: Basic Terms of Disaster Risk Reduction*. International Strategy for Disaster Reduction Secretariat (ed.), Geneva.

ISDR, 2005. *Hyogo Framework for Action 2005–2015: Building the Resilience of Nations and Communities to Disasters*. United Nations International Strategy for Disaster Reduction.

Jaffe, B. E., Borrero, J. C., Prasetya, G. S., Peters, R., McAdoo, B., Gelfenbaum, G., Morton, R., Ruggiero, P., Bretwood, H., Dengler, L., Hidayat, R., Kingsley, E., Kongko, W., Lukijanto, Moore, A., Titov, V. V. and Yulianto, E., 2006. Northwest Sumatra and offshore islands field survey after the December 2004 Indian Ocean tsunami, Great Sumatra Earthquakes and Indian Ocean Tsunamis of December 26, 2004 and March 28, 2005. *Earthquake Spectra*, 22(S3). 105–135.

Jelesnianski, C. P., Chen, J. and Shaffer, W. A., 1992. *Slosh: Sea, Lake, and Overland Surges from Hurricanes*. N. W. Service (ed.), NOAA US Department of Commerce, Washington, DC, 71.

Liu, P. L.-F., Lynett, P., Fernando, H., Jaffee, B. E., Fritz, H., Higman, B., Morton, R., Goff, J. and Synolakis, C., 2005. Observations by the International Tsunami Survey Team in Sri Lanka. *Science*, 308. 1595.

NTIA and Federal Telecommunication Standards Committee, 1996. *Glossary of Telecommunication*

Terms FED-STD-1037B. National Telecommunications and Information Administration (ed.), General Services Administration, Boulder.

Parry, M., Arnell, N., Berry, P., Dodman, D., Fankhauser, S., Hope, C., Kovats, S., Nicholls, R., Satterthwaite, D., Tiffin, R. and Wheeler, T., 2009. *Assessing the Cost of Adaptation to Climate Change: A Review of the UNFCCC and Other Recent Estimates*. International Institute for Environment and Development, Grantham Institute for Climate Change, London.

Penning-Rowsell, E., Johnson, C. and Tunstall, S., 2006. 'Signals' from pre-crisis discourse: lessons from UK flooding for global environmental policy change? *Global Environmental Change*, 16. 323–339.

Synolakis, C. E. and Bernard, E. N., 2006. Tsunami science before and beyond Boxing Day 2004. *Philosophical Transactions of the Royal Society*, 364. 2231–2265.

Synolakis, C. E. and Kong, L., 2006. Runup measurements of the December 2004 Indian Ocean Tsunami. *Earthquake Spectra*, 22(S3). S67–S91.

Synolakis, C. E., Bardet, J.-P., Borrero, J. C., Davies, H. L., Okal, E. A., Silver, E. A., Sweet, S. and Tappin, D. R., 2002. The slump origin of the 1998 Papua New Guinea Tsunami. *Proceedings of the Royal Society*, 458. 763–789.

Synolakis, C., Okal, E. A. and Bernard, E. N., 2005. The Megatsunami of December 26, 2004. *Bridge*, 35(2). 26–35.

Titov, V. V. and Synolakis, C., 1998. Numerical modeling of tidal wave runup. *Journal of Waterway, Port, Ocean, Coast Engineering*, 124. 157–171.

Tripartite Core Group, Government of the Union of Myanmar, United Nations, and ASEAN, 2008. *Post Nargis Joint Assessment Report*. 213.

US Indian Ocean Tsunami Warning System Program, 2007. *How Resilient is Your Coastal Community? A Guide for Evaluating Coastal Community Resilience to Tsunamis and Other Hazards*. US Agency for International Development (ed.), 144.

USAID, 2009. Adapting To Climate Change: A Guidebook For Development Planners. Washington, DC; available at: www.crc.uri.edu/download/CoastalAdaptationGuide.pdf.

Vale, L. J. and Campanella, T. J., 2005. *The Resilient City: How Modern Cities Recover from Disaster*. Oxford University Press, New York.

Walker, B. and Salt, D., 2006. *Resilience Thinking: Sustaining Ecosystems and People in a Changing World*. Island Press, Covelo.

Ward, S. N. and Day, S., 2001. Cumbre Vieja Volcano – Potential collapse and tsunami at La Palma, Canary Islands. *Geophysical Research Letters*, 28. 3397–3400.

research paper

ENVIRONMENTAL HAZARDS
Human and Policy Dimensions

A multi-scale coastal vulnerability index: A tool for coastal managers?

SUZANNE MCLAUGHLIN[1,*] AND J. ANDREW G. COOPER[2]

[1]Land and Property Services NI, Colby House, Stranmillis Court, Malone Lower, Belfast BT9 5BJ, Northern Ireland
[2]School of Environmental Science, University of Ulster, Cromore Road, Coleraine BT52 1SA, Northern Ireland

Coastal hazard management involves the assessment of vulnerability in natural and human environments. Indices incorporating a diversity of indicators have therefore been used extensively to provide spatial analyses of the degree of vulnerability. Such indices are typically applied at global and national scales, and they involve varying degrees of simplification and aggregation of information. The degree of simplification that is desirable depends on the management scale, and higher resolution is required at the local compared to the global scale. To investigate the implications of spatial scale in depicting coastal hazard risk, coastal vulnerability indices were developed at national (Northern Ireland), local authority and site levels. Variables were separated into three sub-indices: a coastal characteristics sub-index concerned with the resilience and susceptibility of the coast to erosion, a coastal forcing sub-index to characterize the forcing variables contributing to wave-induced erosion and a socio-economic sub-index to assess the infrastructure potentially at risk. The three sub-indices were merged to calculate the overall index, which is portrayed in the form of colour-coded vulnerability maps. While a common tripartite index could be employed at national, regional and local scales, the nature of the data used to quantify many of the variables varies according to the scale of management. Some important local variations in vulnerability are masked by simplifications at the national scale. For some variables more detailed information is available as the spatial resolution of the study increases, while others become obsolete as data are of insufficient resolution to differentiate real variability at more detailed scales. The results highlight the importance of spatial scale in developing indices of vulnerability: while a common index architecture can be applied, the selection of variables must take account of the scale at which the hazard is to be assessed. It is likely that limits on index development will also be imposed by data availability at various scales.

Keywords: coastal erosion; coastal management; coastal vulnerability index; geographic information systems-based index; vulnerability; wave-induced erosion

1. Introduction

The concept of coastal vulnerability (Green and McFadden, 2007) is based on human value judgements concerning risk to various elements of the natural and human environment from a variety of sources. The quantification of vulnerability largely depends on the conception of the source of the risk and what element of value to humans is threatened. Since vulnerability is affected by a diverse range of parameters, it is common for geographic information systems (GIS) to be used as a vehicle for integration of data and the creation of indices that express their combined effect (e.g. Thumerer et al., 2000; Bryan et al., 2001). There is widespread concern regarding coastal erosion and its potential impact on coastal communities and infrastructure (Cooper and McKenna, 2008a). In part, this is driven by concerns related to climate and sea-level change (Hinkel and Klein, 2006), and it reflects societal concern over the impacts on human infrastructure and activities rather than the sustainability of the coastal environment itself (Cooper and

■ *Corresponding author. **Email: Suzanne.mclaughlin@dfpni.gov.uk**

doi:10.3763/ehaz.2010.0052 ISSN: 1747-7891 (print), 1878-0059 (online) www.earthscan.co.uk/journals/ehaz earthscan

McKenna, 2008b). Several coastal vulnerability indices have been developed to assess the vulnerability of human activities and infrastructure to sea-level rise (e.g. Gornitz, 1990; Hughes and Brundrit, 1992; Barnett and Adger, 2003) and coastal erosion (e.g. Boruff et al., 2005; Dominguez et al., 2005). Working from the alternative perspective of sustainability of the coastal environment, Pethick and Crooks (2000) developed an index of vulnerability based on geomorphological behaviour that incorporated the frequency of disturbance and relaxation time (post-disturbance recovery).

Although there are different perspectives on the meaning of coastal vulnerability (Klein et al., 1998; Vafeidis et al., 2004; Green and McFadden, 2007), the main purpose of vulnerability assessment is to provide information to guide the process of adaptation and enhance society's adaptive capacity (Kelly and Adger, 2000; Smit and Wandel, 2006). A vulnerability index therefore aims to simplify a number of complex and interacting parameters, represented by diverse data types, to a form that is more readily understood and therefore has greater utility as a management tool.

Vulnerability from particular hazards can be considered at various scales. Past studies of coastal vulnerability using an index approach are most commonly undertaken from global (Gornitz, 1990; Klein and Nicholls, 1999) and regional (Thieler and Hammer-Klose, 2000; Bryan et al., 2001; Dominguez et al., 2005; Mokrech et al., 2008; Dawson et al., 2009) perspectives. Each index is guided by consideration of (i) what is the most appropriate type of data to quantify vulnerability and (ii) what data are available at the spatial scale of interest. Since coastal management is undertaken by organizations operating at various spatial scales, for indices to be useful they must address coastal vulnerability at an appropriate scale. Global-scale approaches enable international approaches to be coordinated and global policies to be debated. At national levels, policy and the prioritization of resources are often based on national-level assessments. The practical response to coastal hazards is, however, most commonly implemented at the local level.

At each scale of management there are different considerations and different types of data. There is no 'one size fits all' index of coastal vulnerability that can be applied at all scales. The components that contribute to vulnerability, the data availability and type, and the utility of an index approach vary with scale. In this paper, therefore, we assess the potential for a multi-scaled coastal vulnerability index based on a common methodological and theoretical framework that has applicability at national, regional and local scales. In this study, these scales relate to national (Northern Ireland), regional (borough council) and local (single beach) levels. The derived indices and the issues that arise in their compilation are discussed.

2. Coastal vulnerability

From a purely anthropocentric perspective concerned with impact on human infrastructure, the vulnerability of a coastal area can be regarded as a function of the physical nature of the coast (which controls its ability to respond to perturbation), the nature (frequency and magnitude) of the perturbation (the forcing factor) and the degree to which such changes impact on human activities or property. Vulnerability can thus be expressed as

$$\begin{aligned}\text{Vulnerability} &\\ &= f \text{ coastal characteristics (resilience} \\ &\quad \text{and susceptibility)} + \text{coastal forcing} \\ &\quad + \text{socio-economic factors}\end{aligned}$$

With an understanding of coastal geomorphology, the coastal forcing and coastal characteristics variables can be used to broadly assess the potential susceptibility of an area to morphological change; however, socio-economic variables are needed to determine the vulnerability from the perspective of the impact of such changes on society. In this regard, a rapidly eroding undeveloped coast, for example, has no vulnerability since it will adjust to perturbation through natural processes and attain a new equilibrium. The socio-economic variables themselves do not, however, quantify all the vulnerability because that depends on the likelihood of physical coastal changes; a stable coast with abundant infrastructure also has low vulnerability. The choice of socio-economic variables adds an inherent cultural bias to the index. Gornitz et al. (1993) suggest that '… it is the perceived social and economic worth of the resources within the region at risk that will determine which, if any, efforts are made to protect a given area…'. Such a purely cost–benefit-based approach is common in coastal defence appraisals (although it often disregards the environmental and social implications; Cooper and McKenna, 2008a). In many coastal vulnerability indices that have previously been developed, socio-economic factors were frequently identified as components that were inadequately incorporated or were recommended for further study (Cooper and McLaughlin, 1998; McLaughlin et al., 2002).

The conceptual basis of the coastal vulnerability index presented here (Figure 1) is that the three elements of physical coastal characteristics, wave forcing and socio-economic characteristics contribute to overall vulnerability. These can be represented in a variety of ways by multiple variables, and the choice of appropriate variables is important in the development of a vulnerability index.

Data for all components were assembled from a variety of digital and analogue sources in a single GIS and a 1 km buffer was employed to extract information for the coastal strip. This enabled assembly of the data to common scales and superimposition of different variables for the calculation of indices.

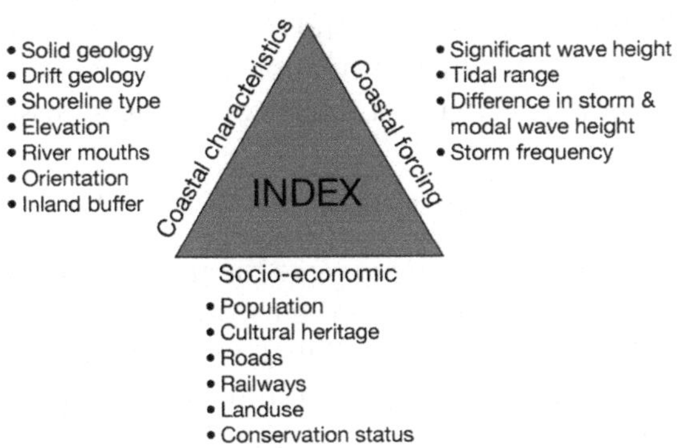

FIGURE 1 Variable classification for sub-indices

3. Variable choices

The number of variables included in published coastal classification indices can vary dramatically. Quelennec (1989) used only three principal variables to identify high-risk coastal areas within Europe whereas Williams et al. (1993) used 54 variables to assess coastal dune vulnerability in Devon and Cornwall. Of the 18 coastal indices reviewed by Cooper and McLaughlin (1998), the majority used between six and 19 variables to calculate various indices.

It is tempting to use all available data in the creation of an index. Dal Cin and Simeoni (1989) claim that '…the more numerous the variables affecting the coast that are taken into consideration then the more correct will be the resulting zoning'. This, however, is not necessarily true since variables can be highly correlated.

In this study, variables that were deemed important from the viewpoint of both scientists and policy makers were used to define (i) the nature of the coast, (ii) the degree to which the coast is exposed to wave energy, and (iii) the principal socio-economic elements for which coastal erosion poses a risk. From a vulnerability perspective, the main consequences of erosion arise from its effects on the human environment. Many elements of this environment have a monetary value related to replacement cost while others (e.g. archaeological monuments or culturally significant sites) are irreplaceable and therefore more difficult to quantify. Cooper and McLaughlin (1998) found that the inclusion of economic factors was the most common recommendation made by authors for improvements to indices. The desirability of including data that reflect one parameter, however, must be balanced against the availability of data. This is particularly problematic in the case of socio-economic variables (McLaughlin et al., 2002).

4. Ranking

The data necessary for compiling a coastal vulnerability index are expressed in qualitative and quantitative form, are frequently available at different scales, and are expressed in different units of measurement. It is thus common practice to assign a rank to each variable to indicate its contribution to vulnerability. In this study, following Gornitz (1990), a scale of 1–5 was chosen, with 5 contributing most strongly to vulnerability and 1 contributing least. The 1–5 scale that was used for every variable standardizes the scoring system and enables variables measured in different units to be combined mathematically.

The ranking of variables is a somewhat subjective exercise, and the criteria by which they are ranked must be clearly defined. We did not weight the different variables relative to each other because, as Cendrero and Fischer (1997, p. 734) contend, 'weighting is difficult because of the diverse number of value judgements that lie behind combined weights'.

The three spatial scales of study selected for investigation (see Figure 2) were the whole of Northern Ireland using $500 \times 500\ m^2$ grid cells (national level), Coleraine Borough Council using $25 \times 25\ m^2$ grid cells (regional level) and Portrush East Strand using $1 \times 1\ m^2$ grid cells (local level) (Figure 3). Although exact boundaries can be placed on administrative areas on maps, natural divisions in the coastal environment do not necessarily coincide with administrative boundaries. Thus, the regional study area also incorporates Benone beach, Magilligan Strand and Runkerry Strand, which although outside the Coleraine Borough Council area, are geomorphologically linked to it. The inland coastal zone boundary for the study at all spatial scales was defined as 1 km inland from the mean high water mark (MHWM).

Tables 1–3 show the variables used at each of the three scales of index and how they were ranked on a 1–5 scale of vulnerability.

Availability of data is an important factor in the selection of parameters to depict vulnerability at each scale. The time taken to process data is also a consideration when data have to be extracted from larger datasets or secondary data are to be derived. For example, selecting and classifying only major roads at the Northern Ireland level from a detailed roads dataset as the scale of the study at the Northern Ireland scale does not permit minor roads to be identified. Full details of how the variables are ranked are given by McLaughlin (2001) and McLaughlin et al. (2002).

4.1. Index calculation

At each scale, the numerical values for the three sub-indices (coastal characteristics, coastal forcing and socio-economic attributes) were summed within the GIS system and the results were then normalized by working the results out as a percentage of the maximum and minimum scores possible. An example of the Northern Ireland coastal characteristics sub-index calculation is shown below:

$$\begin{aligned}\text{Addition of CC vars} &= ([\text{drift geology}] + [\text{elevation}] + [\text{rivers}] \\ &\quad + [\text{inland buffer}] + [\text{geology}] + [\text{orientation}] \\ &\quad + [\text{shoreline type}])\end{aligned}$$

The results of adding the variables for the Northern Ireland scale ranged from 7 to 31 with a possible maximum score of 35 and a minimum of 7. The results of this first calculation were then worked out as a percentage of the range of scores possible as shown below:

$$(([\text{Addition of cc vars}] - 7)/28) \times 100$$

It was then feasible to merge the three sub-indices into an overall index as their scores had been normalized. Gornitz and White (1992) found that an algorithm based on the sum of the variables was less sensitive than one based on the products of the variables. The three sub-indices were therefore

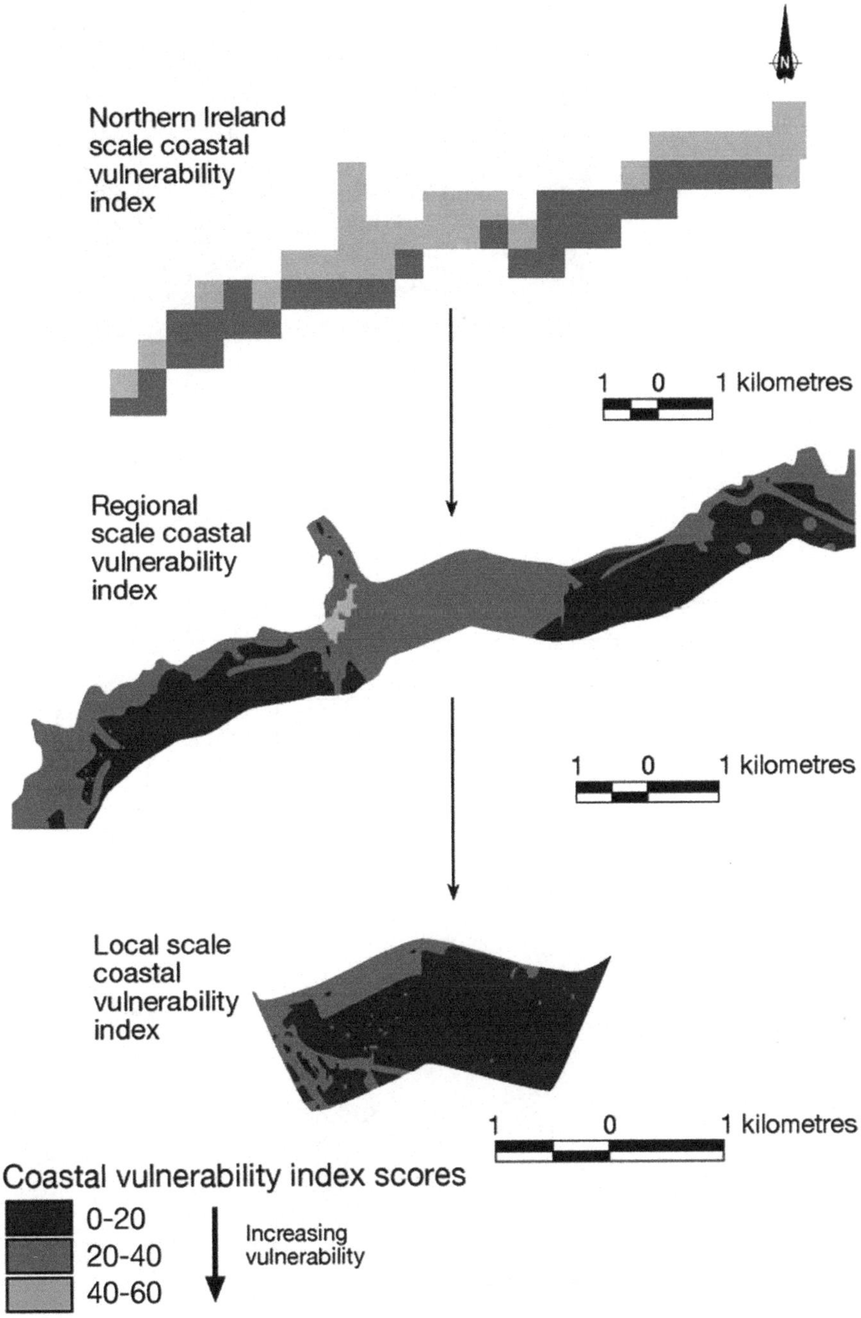

FIGURE 2 Map showing the three spatial scales used in the study: Northern Ireland, Coleraine Borough Council and Portrush East Strand

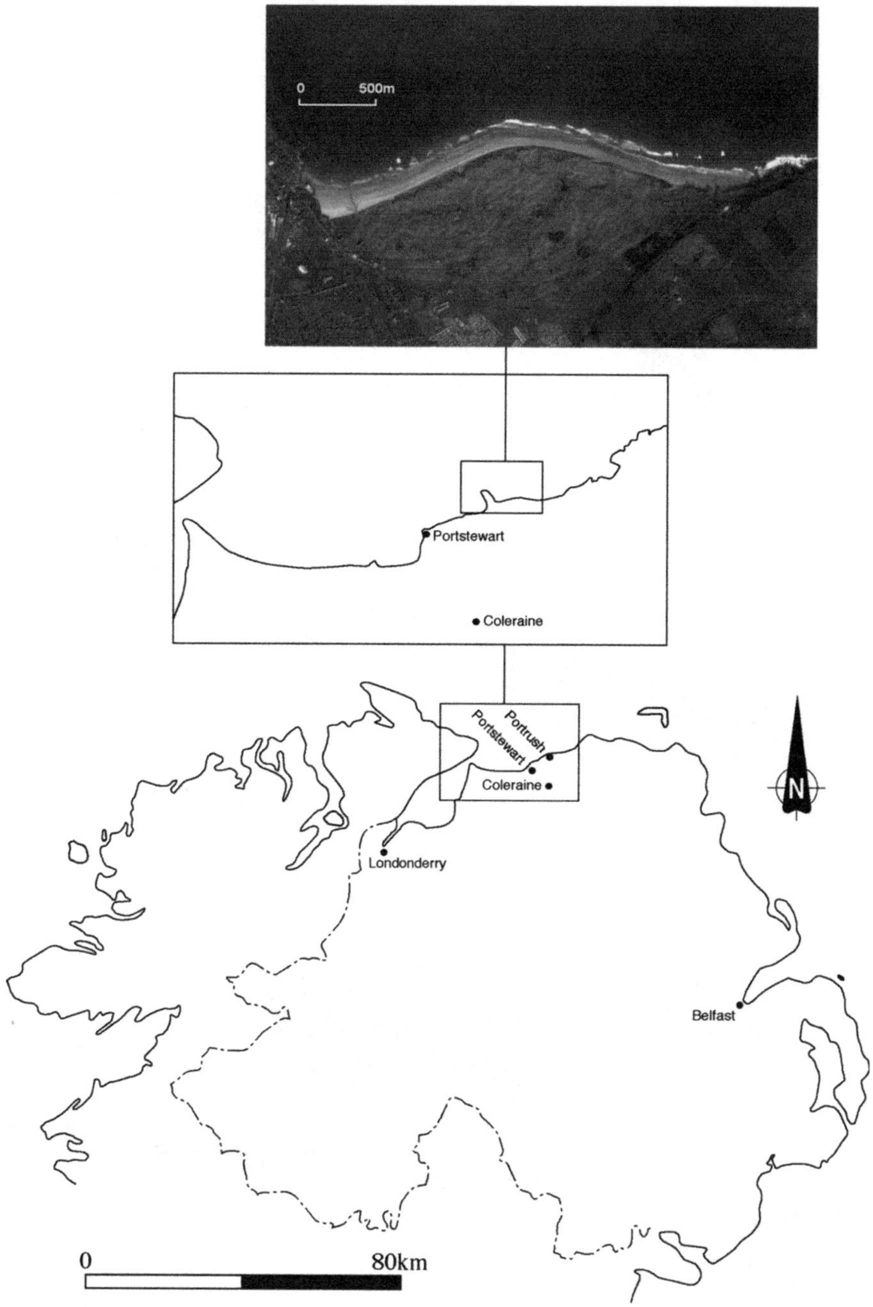

FIGURE 3 Index hierarchy: vulnerability of Portrush East Strand as calculated by each of the index scales (using potential scores). Direct comparisons are not possible as the variables used in each map are not identical, for example the separate *shoreline-type*, *solid geology* and *drift geology* variables of the Northern Ireland scale are replaced by a *landform* variable at the regional scale

Note: Reproduced with kind permission of the controller of HMSO from the 2009 Land & Property Services/Ordnance Survey of Northern Ireland map. Permit No. 90118.

TABLE 1 Northern Ireland vulnerability ranking

	Variable	1	2	3	4	5
Coastal characteristics	Shoreline type	High cliff (40 + m)	Medium cliff (20 to <40 m)	Low cliff (10 to <20 m)	Shingle ridge/ bar	Sand beach/ dune
	Rivers	Absent				Present
	Solid geology	Plutonic Volcanic (lava) High–medium grade metamorphics	Low-grade metamorphics; sandstone and conglomerate well cemented	Most sedimentary rocks	Coarse and/or poorly sorted unconsolidated sediments	Fine unconsolidated sediment, volcanic ash
	Drift geology	Bedrock Urban	Till/boulder clay		Raised beach deposits	Alluvium Blown sand Peat Glacial sands and gravels Glacial outwash sands Recent marine
	Elevation (m)	>30	20 to <30	10 to <20	5 to <10	<5
	Orientation	Not relevant, e.g. sea loughs		Easterly		Northerly
	Inland buffer (m from MHWM)	500 to 1000 m inland				0 to <500 m inland
Coastal forcing	Significant wave height (m)	0 to <0.74 N 0 to <0.24 E	0.74 to <1.49 N 0.24 to <0.48 E	1.49 to <2.23 N 0.48 to <0.72 E	2.23 to <2.98 N 0.72 to <0.96 E	>2.98 N > 0.96 E
	Tidal range (m)	>5	3.5 to <5	2 to <3.5	1 to <2	<1
	Difference in modal and storm waves (m)	<0.10 N <0.10 S	0.10 to <1.70 N 0.10 to <0.25 S	1.70 to <3.30 N 0.25 to <0.40 S	3.30 to <4.90 N 0.40 to <0.55 S	>4.9 N > 0.55 S
	Frequency of onshore storms (%)	0 to <2.8	2.8 to <5.6	5.6 to <8.4	8.4 to <11.2	>11.2
Socio-economic	Settlement	No settlement	Village	Small town	Large town	City
	Cultural heritage	Absent				Present
	Roads	Absent		A-class		Motorway Dual carriageway
	Railways	Absent				Present
	Landuse	Water bodies Marsh/bog and moor Sparsely vegetated areas Bare rocks	Natural grasslands Coastal areas	Forest	Agriculture	Urban and industrial Infrastructure
	Conservation designation	Absent		International		National

TABLE 2 Regional vulnerability ranking

	Variable	1	2	3	4	5
Characteristics	Landform	High resistance cliff	Low resistance cliff	Multiple sand dune ridges	Single dune ridges Gravel and boulder ridges	Mudflat Saltmarsh Beach – no dunes
	Elevation (m)	>30	20 to <30	10 to <20	5 to <10	<5
	Rivers	Absent	Stream	Small river	Medium river	Large river
	Inland buffer (m from the MHWM)	300 to >1000		50 to <300		0 to <50
Forcing	Tidal range (m)	>5	3.5 to <5	2 to <3.5	1 to <2	<1
	Storm probability (based on coastal orientation)	North easterly	Northerly Easterly	North-westerly South-easterly	Southerly South-westerly	Westerly
	Morpho-dynamic state (Dean's parameter)	Rocky coasts and gravel beaches	<1.5 or >5.5	<1.5 to <5.5	>1.5 to >5.5	<1.5 to >5.5
Socio-economic	Cultural heritage	Absent				Present
	Landuse	Water bodies Marsh/bog and moor Sparsely vegetated areas Bare rocks	Natural grasslands Coastal areas	Forest	Agriculture	Urban and industrial Infrastructure
	Population	0 to <5	5 to <20	20 to <50	50 to <100	100 to >200
	Roads	Absent	Minor roads (<4 m)	Minor roads (>4 m)	B-class roads	A-class roads
	Railways	Absent				Present
	Conservation designation	Absent		European International		National

added together and divided by 3 as shown below:

$$\begin{aligned}\text{Coastal vulnerability} &= (\text{coastal characteristics sub-index} \\ &\quad + \text{coastal forcing sub-index} \\ &\quad + \text{socio-economic sub-index})/3\end{aligned}$$

The index was tested via a desktop study to look at historical erosion rates and the presence of coastal defence structures. The results were then portrayed in the form of colour-coded vulnerability maps, enabling the more vulnerable areas to be more easily recognized.

4.2. Scale-specific variables

A greater level of detail is required at the local scale in order to distinguish between areas of potential vulnerability. Thus some variables were included or omitted as the index scale varied. For example, at the Northern Ireland scale, geology was deemed an essential variable to distinguish areas of potential vulnerability to wave-induced erosion. There is a wide variation in the types of both solid and drift geology, with over 40 different categories of solid geology and 10 of drift geology being present in the study

TABLE 3 Local vulnerability ranking

	Variable	1	2	3	4	5
Coastal characteristics	Landform	High resistance cliff Seawall	Low resistance cliff	Multiple sand dune ridges	Single sand dune ridges Gravel and boulder ridges	Mudflat Saltmarsh Beach – no dunes
	Elevation (m)	>30	20 to <30	10 to <20	5 to <10	<5
	Rivers	Absent				Present
	Inland buffer (m from MHWM)	300 to >1000		50 to <300		0 to <50
Coastal forcing	Storm probability (based on coastal orientation)	North-easterly	Northerly Easterly	North-westerly South-easterly	Southerly South-westerly	Westerly
	Morphodynamic state (Dean's parameter)	Inland areas and rocky cliffs	Dissipative or Reflective	Intermediate	Reflective Intermediate Dissipative	Dissipative Reflective
Socio-economic	Cultural heritage	Absent				Present
	Landuse	Rocky cliffs	Scrub	Beach Sand dunes Forest Rough	Agricultural land Tee boxes Fairways Amenity grass	Urban Residential Car parks Greens
	Population	Absent				Present
	Roads	Absent	Footpaths	Minor access roads	B-class roads	A-class roads

area. As the spatial resolution of the study area increases to the local level, this variable becomes obsolete because the data are of insufficient resolution to differentiate vulnerability at a location that comprises only one lithology. At the Northern Ireland scale it is impossible to determine whether it is the solid or drift geology that is the controlling lithology in relation to wave-induced erosion; therefore, a shoreline-type variable was also added, because when used in conjunction with elevation, solid and drift geology it will give a more detailed picture of the actual coastal typology present. A landform variable would have been preferable; however, this was not possible due to a lack of detailed data at this scale.

At the borough council scale however, it is possible to incorporate more detail on the morphology of the coastline compared to the Northern Ireland scale by integrating both the solid and drift geology variables into a more complex and more relevant 'landform' variable. A landform variable has been used by other workers such as Gornitz and Kanciruk (1989), Hughes and Brundrit (1992) and Cambers (1998). Landforms were classified according to their slope, volume and lithology, all of which will determine their vulnerability. For example, a high-resistance cliff will typically have high slope and high volume in comparison with a saltmarsh which has low slope and low volume.

Tidal range and coastal orientation were used at the Northern Ireland and borough council scale to help identify potentially vulnerable areas, but they change so little at the local level that they are not included in that index. The main importance of coastal orientation at the Northern Ireland scale is whether the coast is north or east facing. This determines whether the coast faces into the dominant wind direction for storms, because waves on the north coast are typically much more powerful swell waves compared to the locally generated wind waves of the east coast (Carter, 1982). At the borough council scale, coastal orientation was classified according to 45° compass sectors and used to calculate the directional probability of storms. Hayes's (1979) classification was used to classify tidal range at the Northern Ireland and borough council scale indices. Macrotidal areas were ranked less vulnerable than those with a microtidal regime because the probability of high waves coinciding with high tide and therefore causing erosion is limited by tidal range. In addition, a high tidal range is usually associated with a wide intertidal area in which wave energy can be dissipated. At the local scale tidal range did not vary within the study area and therefore could not be used as a variable to differentiate between vulnerability.

4.3. Scale-independent variables

A number of variables could be used in all three index scales, with the level of detail increasing with the resolution of the study area. For example, elevation data were ranked according to Gornitz and Kanciruk (1989) since elevation is a crucial variable in determining vulnerability to wave-induced erosion. Obtaining increasingly detailed elevation data for the three different scales of index proved difficult. High-resolution elevation data such as Lidar data are expensive and were outside the budget of this study. At the Northern Ireland scale six, 10 m contour datasets were merged together and height data were interpolated for the study area. For the borough council scale it proved difficult to obtain detailed elevation data for areas below 10 m; therefore, the interpolated NI level data were deemed to be of sufficient detail to differentiate elevation at this scale and were used here too. At the local level, three different datasets (one of which included a one-off study funded by the Portrush golf course) were combined to give sufficient data to map the area in detail (height data at a resolution of 1 m). A trade-off was therefore made between the cost of acquisition of elevation data and the level and quality of data used in the study.

This increasing level of detail with the different scales of the indices was also used when incorporating a rivers variable. Rivers were chosen to be included in the indices as river mouths are potentially zones of higher erosion vulnerability due to their generally lower elevation, enabling waves to reach the backshore more easily. Because river mouths can migrate, a buffer of influence was created around them to encompass their region of influence. At the Northern Ireland scale only major rivers were identified, while at the regional scale the influence of smaller rivers becomes of increasing importance and they were consequently incorporated in the index. Data availability was not an issue with this variable as all of the rivers were incorporated in the one detailed dataset with the relevant level of rivers selected for use for each scale.

In relation to the coastal forcing variables, relative exposure to wind, wave and tidal energy largely determines a coast's vulnerability to marine influences. This influence is increased during storms and moderated by tidal range. The distinct decline in wave energy between the high-energy, exposed Atlantic coast in the north and the much lower energy environment of the Irish Sea to the east (Carter and Bartlett, 1990) enabled the Northern Ireland coastline to be split into two parts for development of the coastal forcing sub-index.

An important variable in determining the potential vulnerability of a beach is its modal morphodynamic state as it is closely related to the mobility of the beach (Wright and Short, 1984; Wright et al., 1985). The modal state reflects the modal breaker height and the dominant sediment characteristics. At the Northern Ireland scale index, it was not feasible to break down coastal units according to their spatial variability in breaking wave height, sediment grain size, etc. to use as input to standard equations predicting morphodynamic response. Instead, a generalized parameter regarding the increase in significant wave height (Hs) from modal to storm conditions was used as a representation of potential vulnerability. It was incorporated on the basis that a greater difference between modal and storm waves indicates a greater potential vulnerability; if the energy variations are large, the beach morphology will have to adapt to cope with the conditions, for example through seasonal changes in shape. While many beaches may be in long-term equilibrium, a high degree of seasonal variability represents increased vulnerability from a human perspective.

At the regional level, it was decided to characterize beaches according to Dean's parameter. Dean's parameter Ω (Dean, 1973) can be used to predict the morphodynamic state of the beach, although Jackson et al. (2005) recently showed that it had poor predictive power in Northern Ireland. The modal value of Ω expresses the modal beach state, and the temporal variability of Ω expresses the temporal variability of the beach state (Wright and Short, 1984; Wright et al., 1985). Dean's parameter was therefore calculated for both modal and storm conditions.

All parts of the shoreline, for example, rocky shores and gravel beaches, other than sand beaches were ranked as 1, that is, they have a low vulnerability due to their low to negligible response to modal energy, their very low mobility and their lack of obvious seasonal variability. Reflective and dissipative beaches were ranked as the next least vulnerable forms. This is because they are at the extremes of the spectrum of scores for Dean's parameter, and as such they are normally slow to shift from these modal states (Carter, 1988). Short and Hesp (1982) also noted that, in general, dissipative and reflective beaches have 'very low mobility'. The beaches that move between the dissipative and intermediate states are allocated a vulnerability score of 3, while those that move between the reflective and intermediate states are allocated a score of 4. The reason for this asymmetry is that the range of scores between the reflective and intermediate states is narrower than that between the dissipative and intermediate states, and therefore there is more chance that the beach will respond to changes in dynamic forcing. The final and most vulnerable category is that of beaches that move through all of the beach states from dissipative through intermediate to reflective. These beaches are obviously the most mobile and therefore most vulnerable from a human perspective.

Dean's parameter was included in the local scale index by calculating its value for different sections of the beach to assess beach state. Each section was then classified according to whether it moved between domains or stayed in the same category through both modal and storm conditions. This gives a measure of each sector's propensity to change.

Storms have both economic and geomorphological impacts that vary in time and space. Carter and Bartlett (1990) noted that low-frequency, high-magnitude storms can at times have a much greater impact on the coastline than a series of smaller storms over a few days. Much depends on the recovery period of the beach (Pethick and Crooks, 2000). If the time period between storms is too short for the system to recover, then there is a possibility that a threshold will be reached from which the beach will not recover.

In the index, storm events were defined as winds greater than or equal to 30 knots for at least a 1 h duration with a gap of at least 1 h between storms (MacClenahan et al., 2001). Wind records were used to assess storm frequency and direction. It was decided to add the overall probability (as a percentage) of an onshore storm as a variable to the index.

Identification of onshore storms demands that coastal orientation be established as a preliminary step. In the Northern Ireland scale index, the coast was divided into predominantly north-facing and east-facing coasts. This coastal classification was considered to be too crude for regional- and local-scale assessment and so the coastal orientation was classified into 45° compass sectors.

The identification of areas of coastal vulnerability relies on the physical characteristics of the area along with the marine forcing variables acting on it. However, the extent of the socio-economic environment present will determine to a large degree the coastal management efforts present. Many socio-economic variables have a replacement monetary value while others are irreplaceable and therefore difficult to quantify.

In the socio-economic sub-index, roads were considered as vital lines of communication and the main medium for transport in Northern Ireland. They are also expensive to build, maintain and repair; therefore, erosion of any roads in the coastal area would incur financial costs. For the Northern Ireland index it was decided that sufficient detail could be achieved by including motorways, dual carriageways and A-class roads. At the regional scale it was possible to show more detail; therefore, the vulnerability ranking was adapted to include all of the road classes, including minor roads. At the local scale, data were of a sufficient resolution to include footpaths; this was not possible at either of the two previous scale indices because the resolution was not capable of separately identifying footpaths.

The settlement/population variable was deemed essential to be included in a vulnerability index because it is the perceived 'value' of an area that influences its perceived vulnerability. For example, erosion on an uninhabited island is a process or event rather than damage as there are no negative impacts on human lives. For the Northern Ireland scale index data, various data options were considered. These included the use of census data at ward or enumeration district level. However, as the study area only consists of a 1 km band inland, none of the defined government boundaries are of use. Therefore, for the Northern Ireland scale index it was decided that the identification of conglomerations of settlement such as cities, towns and villages gave a better indication of coastal population. At the regional scale it was deemed feasible to use more detailed census grid point population data. These data were also available in polygon format; however, the grid point data give a much more realistic picture of the population. At the local scale index, the population variable is linked to actual single houses rather than numbers of people per square kilometre. The population data at this scale relate directly to the actual spatial urban form and the physical environment.

Landuse is also a variable that determines the 'value' of the coast and whether it is worth protecting from erosion or not. At the Northern Ireland scale, landuse was identified from the European CORINE landcover classification, which maps a minimum unit of 25 ha^2. Landuse categories were then ranked according to their perceived monetary value. This was also used for the regional level, but at the local scale this was deemed too coarse a resolution; therefore, a customized landuse classification was derived from aerial photography.

With regard to conservation status, there are certain areas that have multiple designations, for example Strangford Lough, which is an AONB (Area of Outstanding Natural Beauty), SPA (Special Protection Area), SAC (Special Area of Conservation), Ramsar site and MNR (Marine Nature Reserve). It was decided, therefore, that for this study the designation status of protected areas in Northern Ireland should be based on whether they were areas of international or national designation. For those with international designations, particularly EU designations, there are direct economic implications if they are not adequately maintained.

National areas were allocated higher vulnerability ratings than international areas as there is arguably a greater chance that their management systems will be weaker, for example, resources are fewer in terms of both expertise and finance. This approach was used for national and regional indices, but was not applicable at the local site.

5. Comparison of the three index scales

To illustrate the comparative resolution and utility of the three indices, the results for a section of the north coast of Northern Ireland are presented using the national-, regional- and local-level indices (Figure 3). To enable relational comparison, the three indices are portrayed using the one legend.

At the Northern Ireland index level, the East Strand at Portrush was classified as being predominantly in the 40–60 vulnerability category. At the regional level this decreases to 20–40, while at the local scale the beach is predominantly in the vulnerability range 0–20. These results show how the scale of study can determine the outcome; an area can be regarded as vulnerable at one scale, but not at another. While the results between the different scales cannot be directly compared, the index values at each scale have been normalized so that each value is relative to the full range of values calculated at that scale. All results are then plotted using the same range of 0–100 to enable relative comparison.

At the Northern Ireland level, Portrush East Strand is being compared with the rest of the coast of Northern Ireland. As it is a soft sandy beach, it is highlighted as comparatively vulnerable in relation to adjacent hard rock cliffs. In relation to the coastal forcing variables, it is an area of high exposure and high energy in contrast to the relatively lower-energy east coast of Northern Ireland. At the regional level, exposure is high throughout the study area, and it is therefore relatively less vulnerable. At the local level, where the variables are being compared only within a single beach, further resolution is obtained and no area of the beach is deemed particularly vulnerable. Although the level of detail increases at the local scale, it has been argued by Goodchild and Quattrochi (1997) that 'generalisation adds information rather than reducing it, because some kinds of geographic information can only become apparent if the world is viewed from a distance'. Similarly, it has been argued that for policy on locally contentious issues such as coastal defence to achieve long-term sustainability, decision making has to be undertaken at the national level, removed from the strong self-interest of local residents (Cooper and McKenna, 2008a).

The results highlight the value of a nested approach to index development and in reinforcing the importance of scale in determining policy response to vulnerability. Coastal zone management operates at a variety of scales within Northern Ireland. The Northern Ireland index gives a broad overview of the relative vulnerability of the coast, giving a visual and spatial perspective to the identification of vulnerable areas. The utility of an index at this scale could be seen, for example, in the implementation of coastal planning restrictions or in the allocation of money to local councils to be spent on coastal defences.

The regional scale index could be used in the development of setback lines or hazard zones to control planning. At the local level, the detail of the index makes it possible to identify infrastructure in potentially highly vulnerable areas or could also be used to devise a local-scale development or management plan for the area. At the local scale, coastal management is a subset of regional management but is also the responsibility of non-government organizations and user groups.

It is therefore up to the coastal manager or policy maker to determine the most appropriate scale or scales of index to apply, depending on whether the policy is a national-, regional- or local-level one, and to also be acutely aware of the potential limitations of the scale or scales of the index they are using.

In an ideal world, indices would be based on local-level information that is aggregated and simplified as larger scales are considered. However, this is not often feasible in terms of data availability, storage and processing, and in terms of the time scale needed to undertake such a study. It is not possible to scale down a macro scale index, as it is impossible to gain further detail from already generalized data. At the Northern Ireland level, for example, detail is therefore sacrificed in order to gain perspective on the vulnerability of the whole coastline. The opposite of this is true at the local level, where the level of detail is high but the overall perspective is low. A local authority's ability to develop an adaptive strategy based on national-level data would be severely constrained as real variations within the local authority area may be masked by simplification. At the national level, in contrast, allocation of resources requires a broad overview of where most work is required in order to develop adaptation strategies in those areas at highest risk.

The three index scales give a choice of the perspective/detail balance. Users can make the trade-off appropriate to their needs by choosing the most suitable index scale for the scale of management. An index at a national level could enable allocation of resources to respond to coastal erosion to those areas with greatest problems, while the regional and local indices could be used to steer utilization of those resources. It must also be noted that the variables within the indices can be adapted as the scales of study change. While some are amenable to this adaptive scaling, others are not. For example, although tidal range is included as a coastal forcing variable at the Northern Ireland and regional scales, it becomes obsolete at the local scale as it is effectively a constant that does not vary substantially along a single beach. Other variables included separately at the Northern Ireland level, such as drift and solid geology, are combined into a landform variable at the regional level. The level of detail necessary to define landforms is now available at this scale, and landform classification is preferred in determining vulnerability. In general, the level of refinement is increased as the resolution of the index is increased. For example, the classification of roads has sufficient detail to include footpaths at the local scale. While settlements were ranked as cities, towns or villages at the Northern Ireland level, they are classed at the level of individual houses at the local scale. Modelled wave parameters are obtained from bathymetric and model output grids whose resolution improves towards the local scale.

6. Discussion

Coastal vulnerability indices have been created for different purposes and the vulnerability according to what risk is being considered. They, of necessity, rely on what data are available rather than what might be desired in an ideal world. In this study, we were able to assimilate data at different spatial scales and adapt them to portray coastal vulnerability in a common format at three different scales. This study indicates that spatial scale is also an important consideration in the development of coastal vulnerability indices. No single scale is suitable for all needs. Different scales tend to reflect different priorities, and the influence of a given variable will increase or decrease as scale changes. As Bian (1997) noted, 'one of the central arguments about scale is that data may carry different information when presented at different scales'. This is true in relation to coastal environments as processes acting at various scales condition different aspects of the coast. The temporal limitations of using indices must also be noted as the variables are static pictures of a single period in time.

Multi-scaled indices like those presented here have the capacity to be used in assisting with devising coastal policy in the face of various coastal hazards. The presentation of information at a scale meaningful to the management unit enhances the utility of that information for development of adaptation strategies.

References

Barnett, J. and Adger, W. N., 2003. Climate dangers and atoll countries. *Climatic Change*, 61. 321–337.

Bian, L., 1997. Multiscale nature of spatial data in scaling up environmental models. *Scale in Remote Sensing and GIS*, M. F. Goodchild and D. A. Quattrochi (eds). Lewis Publishers, London. 13–26.

Boruff, B. J., Emrich, C. and Cutter, S. L., 2005. Erosion hazard vulnerability of US coastal counties. *Journal of Coastal Research*, 21. 932–942.

Bryan, B., Harvey, N., Belperio, T. and Bourman, B., 2001. Distributed process modeling for regional assessment of coastal vulnerability to sea-level rise. *Environmental Modeling and Assessment*, 6. 57–65.

Cambers, G., 1998. *Coping With Beach Erosion*. UNESCO Publishing, Paris.

Carter, R. W. G., 1982. The coast. *Northern Ireland Environment and Natural Resources*, J. G. Cruickshank and D. N. Wilcock (eds). Queen's University of Belfast, Belfast. 119–138.

Carter, R. W. G., 1988. *Coastal Environments*. Academic Press, London.

Carter, R. W. G. and Bartlett, D. J., 1990. Coastal erosion in northeast Ireland – Part 1: sand beaches, dunes and river mouths. *Irish Geography*, 23. 1–16.

Cendrero, A. and Fischer, D. W., 1997. A procedure for assessing the environmental quality of coastal areas for planning and management. *Journal of Coastal Research*, 13. 732–744.

Cooper, J. A. G. and McKenna, J., 2008a. Social justice and coastal erosion management: the temporal and spatial dimensions. *Geoforum*, 39. 294–306.

Cooper, J. A. G. and McKenna, J., 2008b. Working with natural processes: the challenge for coastal protection strategies. *Geographical Journal*, 174. 315–331.

Cooper, J. A. G. and McLaughlin, S., 1998. Contemporary multidisciplinary approaches to coastal classification and environmental risk analysis. *Journal of Coastal Research*, 14(2). 512–524.

Dal Cin, R. and Simeoni, U., 1989. Coastal zoning and vulnerability: application to the Middle Adriatic (Italy). *Coastal Zone '89*. 98–110.

Dawson, R. J., Dickson, M., Nicholls, R. J., Hall, J., Walkden, M. J. A., Stansby, P. K., Mokrech, M., Richards, J., Zhou, J., Milligan, J., Jordan, A., Pearson, S., Rees, J., Bates, P. D., Koukoulas, S. and Watkinson, A., 2009. Integrated analysis of risks of coastal flooding and cliff erosion under scenarios of long term change. *Climatic Change*, 95(1–2). 249–288.

Dean, R. G., 1973. Heuristic models of sand transport in the surf zone. *Proceedings of the 1st Australian Conference on Coastal Engineering*, 208–214, Sydney, Australia.

Dominguez, L., Anfuso, G. and Gracia, F. J., 2005. Vulnerability assessment of a retreating coast in SW Spain. *Environmental Geology*, 47. 1037–1044.

Goodchild, M. F. and Quattrochi, D. A., 1997. Scale, multiscaling, remote sensing, and GIS. *Scale in Remote Sensing and GIS*, D. A. Quattrochi and M. F. Goodchild (eds). Lewis Publishers, London. 1–11.

Gornitz, V., 1990. Vulnerability of the East Coast, USA to future sea level rise. *Journal of Coastal Research*, 1(Special Issue No. 9). 201–237.

Gornitz, V. and Kanciruk, P., 1989. Assessment of global coastal hazards from sea level rise. *Coastal Zone '89*. 1345–1359.

Gornitz, V. M. and White, T. W., 1992. *A Coastal Hazards Database for the U.S. East Coast*. Report to the US Department of Energy, 3913, Oak Ridge National Laboratory, TN, USA.

Gornitz, V., Daniels, R. C., White, T. W. and Birdwell, K. R., 1993. *The Development of a Coastal Risk Assessment Database: Vulnerability to Sea-Level Rise in the U.S. Southeast*. DE-AC05-84OR21400, US Government Report, Oak Ridge National Laboratory, TN, USA.

Green, C. and McFadden, L., 2007. Coastal vulnerability as discourse about meanings and values. *Journal of Risk Research*, 10. 1027–1045.

Hayes, M. O., 1979. Barrier island morphology as a function of tidal and wave regimes. *Barrier Islands*, S. P. Leatherman (ed.). Academic Press, New York. 1–27.

Hinkel, J. and Klein, R. J. T., 2006. Integrating knowledge for assessing coastal vulnerability to climate change. *Managing Coastal Vulnerability: An Integrated Approach*, L. McFadden, R. J. Nicholls and E. C. Penning-Rowsell (eds). Elsevier Science, Amsterdam, The Netherlands. 61–77.

Hughes, P. and Brundrit, G. B., 1992. An index to assess South Africa's vulnerability to sea-level rise. *South African Journal of Science*, 88(June 1992). 308–311.

Jackson, D. W. T., Cooper, J. A. G. and Del Rio, L., 2005. Geological constraints on beach morphodynamic state. *Marine Geology*, 216. 297–314.

Kelly, P. M. and Adger, W. N., 2000. Theory and practice in assessing vulnerability to climate change and facilitating adaptation. *Climate Change*, 47. 325–352.

Klein, R. J. T. and Nicholls, R. J., 1999. Assessment of coastal vulnerability to climate change. *Ambio*, 28. 2182–2187.

Klein, R. J. T., Smit, M. J., Goosen, H. and Hulsbergen, C. H., 1998. Resilience and vulnerability: coastal dynamics or Dutch dikes? *Geographical Journal*, 164. 259–268.

MacClenahan, P., McKenna, J., Cooper, J. A. G. and O'Kane, B., 2001. Identification of highest magnitude coastal storm events based on an analysis of wind speed and duration thresholds in western Ireland. *International Journal of Climatology*, 21. 829–842.

McLaughlin, S., 2001. Assessment and development of a coastal vulnerability index for Northern Ireland employing GIS techniques. Unpublished Ph.D. thesis. University of Ulster at Coleraine.

McLaughlin, S., McKenna, J. and Cooper, J. A. G., 2002. Socio-economic data in coastal vulnerability indices: constraints and opportunities. *Journal of Coastal Research*, 36. 487–497.

Mokrech, M., Nicholls, R. J., Richards, J. A., Henriques, C., Holman, I. P. and Shackley, S., 2008. Regional impact assessment of flooding under future climate and socio-economic scenarios for East Anglia and North West England. *Climatic Change*, 90. 31–55.

Pethick, J. and Crooks, S., 2000. Development of a coastal vulnerability index: a geomorphological perspective. *Environmental Conservation*, 27. 359–367.

Quelennec, R. E., 1989. The Corine 'Coastal Erosion Project': identification of coastal erosion problems and data base on the littoral environment of eleven European countries. *Coastal Zone '89*, 4594–4601.

Short, A. D. and Hesp, P. A., 1982. Wave, beach and dune interactions in southeastern Australia. *Marine Geology*, 48. 259–284.

Smit, B. and Wandel, J., 2006. Adaptation, adaptive capacity and vulnerability. *Global Environmental Change*, 16. 282–292.

Thieler, E. R. and Hammer-Klose, E. S., 2000. *National Assessment of Coastal Vulnerability to Sea-Level Rise: Preliminary Results for the US Pacific Coast.* Open File Report 00–178. 1, United States Geological Survey (USGS), Woods Hole, MA, USA.

Thumerer, T., Jones, A. P. and Brown, D., 2000. A GIS based coastal management system for climate change associated flood risk assessment on the east coast of England. *International Journal of Geographical Information Science*, 14. 265–281.

Vafeidis, A. T., Nicholls, R. J., McFadden, L., Hinkel, J. and Grasshoff, P. S., 2004. Developing a global database for coastal vulnerability analysis: design issues and challenges. *The International Archives of Photogrammetry, Remote Sensing and Spatial Information Sciences*, 35:B. 801–805.

Williams, A. T., Davies, P., Curr, R., Koh, A., Bodere, J., Hallegouet, B., Meur, C. and Yoni, C., 1993. A checklist assessment of dune vulnerability and protection in Devon and Cornwall, UK. *Coastal Zone '93*, 3394–3408.

Wright, L. D. and Short, A. D., 1984. Morphodynamic variability of surf zones and beaches: a synthesis. *Marine Geology*, 56. 93–118.

Wright, L. D., Short, A. D. and Green, M. O., 1985. Short-term changes in the morphodynamic states of beaches and surf zones: an empirical predictive model. *Marine Geology*, 62. 339–364.

research paper

ENVIRONMENTAL HAZARDS
Human and Policy Dimensions

Resilience in the context of tsunami early warning systems and community disaster preparedness in the Indian Ocean Region

FRANK THOMALLA[1,2,*] AND RASMUS KLOCKER LARSEN[2]

[1]Department of Environment and Geography, Macquarie University, NSW 2109, Australia
[2]Stockholm Environment Institute (SEI), Kräftriket 2B, Stockholm SE 106 91, Sweden

This paper discusses insights from post-tsunami early warning system (EWS) development in Thailand, Sri Lanka and Indonesia by analysing selected elements of resilience, based on the Coastal Community Resilience (CCR) framework, and by distinguishing between the cognitive, normative and procedural dimensions of EWSs. The findings indicate that (1) recent calls to develop participatory and people-centred EWSs as promoted by the Hyogo Framework for Action 2005 – 2015 have not been sufficiently translated into action in the implementation of national policies and strategies for early warning; (2) policy and guidance places significantly more emphasis on the procedural compared to the normative and cognitive dimensions of EWSs; (3) practitioners engaged in early warning and disaster risk reduction operate in contexts shaped by multiple stakeholder agendas and face considerable challenges in negotiating diverse needs and priorities; and (4) few platforms currently exist that enable stakeholders to coordinate and reconcile agendas, negotiate joint targets, share knowledge and critically reflect on lessons learnt, and to improve the integration of early warning with other priorities such as livelihoods improvement, natural resource management and community development.

Keywords: community-based disaster risk management; early warning system; Indian Ocean; last mile; resilience

1. Introduction: Linking early warning with communities at risk

Following the 2004 tsunami, the development of the Indian Ocean Tsunami Warning and Mitigation System was initiated at the World Conference for Disaster Reduction in 2005 under the lead of the United Nations Education Scientific and Cultural Organization's Intergovernmental Oceanographic Commission. Initial consultations undertaken by the Stockholm Environment Institute (SEI) with stakeholders in 2008 emphasized that a large number of organizations were engaged in early warning system (EWS) development and community-based disaster risk management (CBDRM) in the region and that considerable knowledge regarding community linkages of the EWS existed. However, concerns were voiced that the technological aspects of EWS development had been receiving considerably more attention than human aspects such as hazard awareness, disaster preparedness, reconciling priorities in the context of multiple agendas, and motivation and support for CBDRM activities that link with early warning efforts. Even though the importance of addressing community linkages in EWSs is strongly emphasized in current guidance, practitioners face considerable challenges in applying these insights in their operational contexts. Based on these findings, SEI, together with regional partners, conducted a multi-stakeholder participatory assessment to provide a platform for EWS stakeholders, to create an improved understanding of the challenges and enabling conditions for practitioners to implement recommendations and guidance.

In hazard early warning systems (EWSs), the term 'last mile' or 'last kilometre' is frequently

■ *Corresponding author. *Email:* frank.thomalla@mq.edu.au

doi:10.3763/ehaz.2010.0051 © 2010 Earthscan ISSN: 1747-7891 (print), 1878-0059 (online) www.earthscan.co.uk/journals/ehaz earthscan

used to describe the dissemination of a warning of an impending hazard to the largest number of people at risk in the shortest possible time. The development and implementation of effective systems and procedures for the delivery of credible warnings is typically seen as a considerable challenge – so much so that some actors refer to the 'last 99 miles' (e.g. Hollister, 2008, personal communication). The term 'last mile' has been criticized by some researchers (e.g. Richardson and Paisley, 1998; Twigg, 2003; Lassa, 2008; Kelman, 2009) because it refers to an approach in which the delivery of warnings to the people at risk is the last step in a top-down approach to EWS development. In their view, EWSs should be developed using a bottom-up approach that prioritizes the needs of the communities/users and engages them more actively in the development of the system. In this approach, the interface of the community with the EWS is seen as the 'first mile'. Others regard this distinction as semantics. Some actors refer to integrated 'end-to-end' or 'people-centred' EWSs that address all stages of early warning, from hazard detection and warning to community-level response (US-IOTWS, 2007).

The notion of the last mile has been popularized in the Indian Ocean in relation to tsunami EWS development. As a notion, it is understood and interpreted differently by different actors. For example, LIRNE Asia in Sri Lanka regards the last mile as a challenge for rural communities to access media and address this by supplementing traditional media channels for warning dissemination with additional technologies (LIRNE Asia, 2008). The US Indian Ocean Tsunami Early Warning System (IOTWS) distinguishes between 'upstream' and 'downstream' components of the EWS and identifies education, mitigation efforts, identification of safe areas and development of local decision-making procedures as priorities for 'last mile' assistance (US-IOTWS, 2007). Singh Bedi (2006) understands the 'last mile' as the capacity of the community to take action in response to a received warning and therefore supports the development of the capacities of local institutions. From the perspective of policy implementation, it is not relevant to seek a universal definition of the first mile/last mile. Rather, the diversity in interpretations hints at the complexities associated with the links between technology and communities in the development of national EWSs in the Indian Ocean Region and the current challenges associated with improving this link.

After the 2004 tsunami, the United Nations Educational, Scientific, and Cultural Organization's (UNESCO) Intergovernmental Oceanographic Commission (IOC) received a mandate from the international community to coordinate the establishment of the IOTWS at the World Conference on Disaster Reduction (WCDR) in 2005. The *Hyogo Framework for Action 2005–2015* (HFA) was adopted at the WCDR and recognizes early warning as an effective tool to reduce vulnerabilities, save lives and help protect livelihoods and national development gains, and to improve preparedness and response to natural hazards. The HFA takes on the more critical perspective of the 'last mile' in stressing that disaster risk reduction (DRR) must be 'underpinned by a more pro-active approach to informing, motivating and involving people in all aspects of disaster risk reduction in their own local communities' through multi-stakeholder and cross-sectoral partnerships (UN/ISDR, 2005, p. 2). One participant concluded that a core message from the WCDR was that 'to be effective early warning systems must be embedded in, understandable by, and relevant to the communities which they serve' (Moench, 2005).

The WCDR also saw the launch of the International Early Warning Programme (IEWP) and the formation of the Intergovernmental Coordination Group (ICG). The ICG was formed under the auspices of the IOC to serve as the regional body to plan and coordinate the design and implementation of the IOTWS. The ICG initially established four working groups at the First Session of the ICG/IOTWS tasked with developing the technical plans for the warning system. The discussions of this meeting focused on technology transfer, and two additional working groups, one on risk assessment and the other on mitigation, preparedness and response, were established at the Second and Third Sessions of the ICG/IOTWS, respectively. In 2005 the United Nations Secretary-General requested a global survey of EWSs with a view to advancing the development of a global EWS for all natural hazards (UN/ISDR, 2006a). The report concluded that while some warning systems were well advanced, there were numerous gaps and shortcomings, especially in developing countries and in terms of effectively reaching and serving the needs of those at risk. Also in 2005, national assessments of 16 Indian Ocean countries were conducted to identify capacity-building needs and support requirements for developing an IOTWS (www.ioc-tsunami.org). The assessments investigated the legal state of affairs and national institutional structures, but did not address issues relating to the 'last mile' (IOC et al., 2005).

As part of the international response to the 2004 tsunami, the US Agency for International Development launched the US-IOTWS Program in August 2005 (US-IOTWS, 2008). This two-year programme aimed to provide technical assistance to the region through an integrated 'end-to-end' approach that addressed all aspects of EWS development from hazard detection and warning to community-level response (US-IOTWS, 2008). The programme included regional, national and local as well as cross-cutting activities encompassing all 28 Indian Ocean countries but focused in particular on Indonesia, Sri Lanka, India, Thailand and the Maldives. The Third International Conference on Early Warning in 2006 emphasized the need to strengthen the IEWP and the Platform for the Promotion of Early Warning (PPEW, 2009) as part of the United Nations International Strategy for Disaster Reduction (UN/ISDR) system, and to develop concrete measures and project ideas to implement the HFA (UN/ISDR, 2006b). A key outcome was the establishment of the Indian Ocean Consortium, a multi-agency consortium that aims to support the development of national components of the IOTWS by fostering coordination mechanisms among governments, supporting implementation of national plans for tsunami warning, and creating linkages between regional efforts (IOC, 2008). In 2007, 25 out of 28 IOTWS participating countries had established official focal points for disseminating warning information (UN/ISDR, 2007b). In December 2007, the US Government and UNESCO/IOC co-sponsored a forum in Bangkok, Thailand, to review progress and define priorities for future development and sustainability of the IOTWS with national and regional partners at a transition workshop (US-IOTWS, 2008).

Enquiries undertaken by the Stockholm Environment Institute (SEI) and partners in 2008 with regional stakeholders, including the UN/ISDR, the United Nations Development Programme Regional Centre for Asia and the Pacific and the Asian Disaster Preparedness Centre (ADPC), indicated that despite the actions taken so far, there was among policy makers and practitioners at international and regional levels a widespread sense of a lack of implementation on the 'last mile' and the mainstreaming of DRR as promoted under the HFA. Following the typology of Smith (2005), there has been significant investment in two of the three interrelated stages of the early warning process, namely evaluation/forecasting (the scientific and technical dimension) and warning/dissemination (the institutional and political dimension), but little attention to the response (the human dimensions of risk perception and decision making) (see also Hamza, 2006). In 2006, 20 of the 28 IOTWS participating countries lacked response plans for early warning (UN/ISDR, 2006b, 2007b). UN/ISDR stated at that time that 'what needs to be done to address the shortcomings is not a mystery, but has been already laid out in general terms in a succession of documents and meetings over the last decade' (UN/ISDR, 2006a, p. vi). The ICG/IOTWS Secretariat noted that 'for all the progress and improvement at the detection end of the system, the hard reality is that much remains to be done to ensure dissemination of effective warnings and improve the preparedness of communities to respond to such warnings' (Elliot, 2006, p. 5).

The aim of this paper is to explore the links between technology and communities in the development of national EWSs in the Indian Ocean Region. Departing from the challenges associated with implementing policy for EWSs on the 'last mile', it aims to illustrate how the current challenges of implementing the HFA and national action plans derive from the fact that the critical perspective on the 'last mile' espoused in the HFA is not translated into action.

2. Methodology

2.1. Framework for examining EWS policy implementation

While the end users of early warnings have been subject to a great deal of research, very few systemic enquiries have been conducted into the divergent experiences and priorities of actors in the entire warning chain and how these influence policy implementation. In this paper we depart from the appreciation that early warning and DRR address 'over-determined' problems, that is, problems that are characterized by multiple legitimate perspectives on what constitutes the actual causes of a given risk (Powell and Jiggins, 2003). Because 'outsiders' and 'insiders' measure and describe risk in very different ways (Salter, 1996; Twigg, 2003), in over-determined problem situations, the distinction between risk perception and actual risk loses its justification (Beck, 1992). Just as people in a warning situation do not always respond rationally to warnings (e.g. Buchanan-Smith and Davies, 1995; Thomalla and Schmuck, 2004), EWS implementation is not an entirely rational and logical process. Rather, it consists of a wide range of social and organizational processes that employ technological means to reduce risks and losses (Hamza, 2006). Warning channels can thus be described as chaotic patchworks of communication (Andersen, 2007) that require multiple iterative coordinative actions between agencies, officials and citizens (Rego, 2001; De Marchi, 2007).

The challenge of implementation suggests that existing policy and guidance for early warning cannot fulfil its role if it is treated as a knowledge prescriptive instrument. Instead, it has to be understood as a process that builds principles for action for communities of practice, creating a 'space of meaning' with theories for action, social change and instruments for implementation (SLIM, 2004). Because each operational context is unique, stakeholders who aim to implement a policy or strategy have to learn their way into this implementation, often with a considerable need for innovation. This places implementation of recommended actions in a 'community of practice', a group of stakeholders who are interacting regularly in a certain manner and with a certain set of values, assumptions and actions (Wenger, 1998).

The approach outlined by the US-IOTWS guide for Coastal Community Resilience (CCR) (US-IOTWS, 2007) can be used to illustrate this

TABLE 1 Resilience elements and desired outcomes (US-IOTWS, 2007)

A. *Governance:* Leadership, legal framework and institutions provide enabling conditions for resilience through community involvement with government
B. *Society and economy:* Communities are engaged in diverse and environmentally sustainable livelihoods resistant to hazards
C. *Coastal resource management:* Active management of coastal resources sustains environmental services and livelihoods and reduces risks from coastal hazards
D. *Land use and structural design:* Effective land use and structural design that complement environmental, economic and community goals and reduce risks from hazards
E. *Risk knowledge:* Leadership and community members are aware of hazards and risk information is utilized when making decisions
F. *Warning and evacuation:* The community is capable of receiving notifications and alerts of coastal hazards, warning at-risk populations and individuals acting on the alert
G. *Emergency response:* Mechanisms and networks are established and maintained to respond quickly to coastal disasters and address emergency needs at the community level
H. *Disaster recovery:* Plans are in place prior to hazard events that accelerate disaster recovery, engage communities in the recovery process, and minimize negative environmental, social and economic impacts

point. The framework, developed in a series of workshops with government agencies, non-governmental organizations (NGOs) and individuals, comprises eight elements essential for CCR (Table 1). These elements of resilience incorporate long-term planning and implementation, hazard event-oriented resilience elements and governance, which provide the enabling framework for resilience in all other elements. Risk knowledge is similarly a cross-cutting requirement within each resilience element (US-IOTWS, 2007).

In the CCR, each resilience element is assessed with reference to a benchmark that represents the desired conditions against which the resilience status of a coastal community is evaluated (US-IOTWS, 2007, pp. 3–4). The resilience of the community is evaluated by using selected rating systems to rank the assessment results against these benchmarks, to position the current situation of the community and to suggest plans and means for action.

When investigating policy and guidance for EWS development as a process in which knowledge is jointly constructed by the stakeholders involved, three dimensions can be discerned: cognitive, normative and procedural (SLIM, 2004). These can be expressed in the language of the CCR framework as follows (Figure 1):

1. *Normative dimension*: The identification of the resilience benchmark implies an implicit theory of social change through which to determine what qualities and levels are to be considered desirable for each element.
2. *Cognitive dimension*: The rating of current levels of resilience against the desired and the creation of plans for action depends on the cognitive styles of those involved.
3. *Procedural dimension*: The development and application of means and instruments to move from the current level of resilience

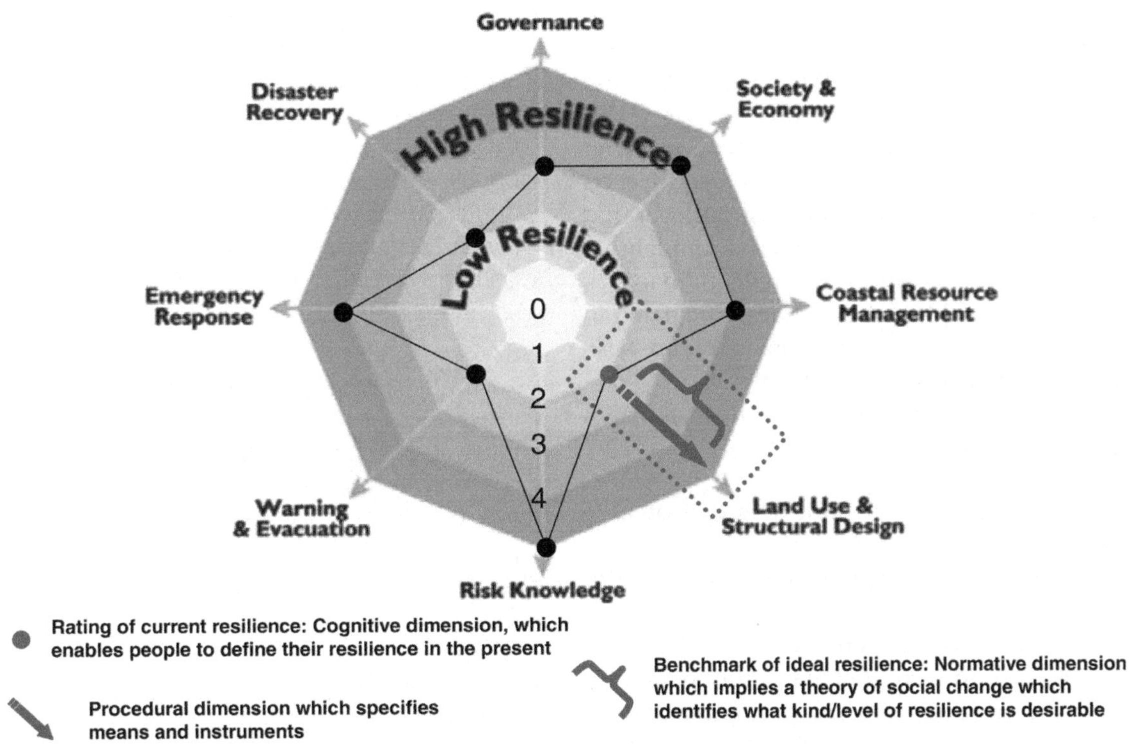

FIGURE 1 Role of normative, cognitive and procedural dimensions of EWS policy, exemplified with the CCR framework
Source: US-IOTWS, 2007.

towards the desired one are embedded in a process of interaction between stakeholders.

2.2. Stakeholder consultations of last mile implementation experiences

A participatory assessment was conducted in 2008 to provide a platform for stakeholders to jointly create an improved understanding of the challenges and enabling conditions for different actors to implement recommendations and guidance to strengthen technology – community linkages of EWSs (the 'last mile'). In June 2008, organizations from the region were invited to participate in the assessment through email communication. SEI, in collaboration with ADPC and Raks Thai Foundation (RTF), undertook a series of stakeholder consultations in Thailand, Sri Lanka and Indonesia between July and December 2008. Consultations were held with government agencies involved in disaster management, coastal resource management and community development at different administrative levels, international and national NGOs and community-based organizations (CBOs), and communities at risk from tsunamis and other coastal hazards (Table 2) in order to elicit factors at the national, provincial and district level that contribute to or limit the effective implementation of early warning. This reflects many of the same stakeholder groups that participated in the CCR framework development, and also more local groups and communities.

The following questions guided the research process:

- What is the state of understanding of EWS community linkages ('last mile')?
- Why are identified issues for the last mile not being addressed?
- Which conditions enable/hinder stakeholders to implement recommendations in practice?
- Which learning process are organizations going through to be able to implement policy and practice recommendations?

TABLE 2 Participating agencies, organizations and communities in selected case studies in Sri Lanka, Thailand and Indonesia

Sri Lanka

- Disaster Management Centre (DMC), Ministry of Disaster Management & Human Rights
- Ministry of Fisheries and Aquatic Resources
- Coastal Conservation Department (CCD)
- Department of Meteorology
- Sri Lanka Red Cross (SLRC)
- UNDP Disaster Management Program
- IUCN – The World Conservation Union, Sri Lanka Country Office
- Practical Action
- LIRNE Asia
- Sewalanka Foundation
- Mawella village of Hambantota District

Krabi Province, Thailand

- Thai Department of Disaster Prevention and Mitigation
- Krabi Provincial Administration Division of Public Works
- Krabi Provincial Administration Organization (PAO)
- Raks Thai Foundation
- Thai Red Cross
- Save the Andaman Network (SAN)
- Monitoring, Control and Surveillance (MCS) fisheries network
- Emergency Service (EMS)
- One Tambon One Security (OTOS)
- CBDRM committee, village Ban Klong Prasong
- CBDRM committee, village Ban ThaKlong
- CBDRM committee, village Ban Thalane
- Rescue Team for marine accident management, Koh Punyee
- CBDRM committee, village Ban Nam Khem

Indonesia

Padang

- PUSDALOPS (Provincial 24/7 Emergency Operations Centre)
- Komunitas Siaga Tsunami (KOGAMI)
- Centre for Disaster Studies, Andalas University
- BAPPEDA (Municipal Government Planning Board for Padang)
- Municipal Government

Banda Aceh

- IFRC
- German Red Cross
- American Red Cross
- Irish Red Cross

Continued

TABLE 2 Continued

• GTZ Aceh Rehabilitation and Reconstruction Programme
• BRR (Agency for Rehabilitation and Reconstruction for Aceh and Nias)
• Sea Defence Consultants
• Tsunami and Disaster Mitigation Research Centre (TDMRC)
Jakarta
• UNDP Crisis Prevention and Recovery Unit (CPRU)
• Indonesian Red Cross (Palang Merah Indonesia – PMI), Aceh Darussalam Chapter
• Indonesian Institute of Sciences (Lembaga Ilmu Pengetahuan Indonesia – LIPI)
• GTZ International Services
• Badan Nasional Penanggulangan Bencana (BNPB)
• Badan Meteorologi dan Geofisika (BMG), ASEAN Earthquake Information Centre, Earthquake Engineering and Tsunami Division
• State Ministry of Research and Technology (RISTEK)
• Information Centre for Research on Natural Disasters (PIRBA)
• UNESCO Indonesia, Jakarta Tsunami Early Warning Centre (JTIC)
• BGR
• UNDP
• GITEWS Early Warning & Mitigation Centre

To understand how organizations seek to implement policy in their specific operational context, the consultations were guided by a methodology of investigating innovation histories (Douthwaite and Ashby, 2005). This tool enables people who have been involved in a learning process to record and reflect on how this learning took place. An adapted version of the policy life cycle of Parsons (1995, cf. Lindahl, 2008) was used as a dialogical tool in discussions with stakeholders to explore the guidance and policy related to the 'last mile'. Case studies for detailed analysis were identified and selected based on the following criteria:

- The selected cases were highly exposed to coastal hazards.
- There was a dominant sense among decision makers of the urgency in establishing an EWS.
- The actors and initiatives placed a significant emphasis on issues relating to early warning and disaster preparedness.
- The selected cases addressed multiple hazards.
- The selected cases represented operations at different scales and/or different organizational entry points for the project team.
- The selected cases exhibited different levels of hazard awareness and different levels of previous disaster preparedness experiences.

3. Key findings

In this section, we discuss the evidence from the stakeholder consultations, substantiated with secondary data, in terms of the normative, cognitive and procedural challenges associated with EWS implementation. Three selected CCR elements and benchmarks are used as points of departure for this discussion, namely the two cross-cutting elements governance and risk knowledge, and one that focuses on the concrete aspects of warning (Table 3). Each section below therefore starts with a discussion of the normative challenges associated with the often very diverse views on what, for each stakeholder, comprise desirable qualities of resilience.

3.1. Governance

3.1.1. Normative: Trajectories of social change

Disasters can affect everyone and the HFA describes them as 'everybody's business'. They unite stakeholders despite other diverging interests. However, EWS plans and actions have a bearing on and are themselves affected by longer trajectories of social change. The implementation of EWSs after the 2004 tsunami has taken place in a contested legal environment. For example, in Banda Aceh, Indonesia, the peace process and the tremendous scale of the reconstruction efforts shape the understanding of EWSs. The tsunami has also led to decimation in government capacity. In Sri Lanka the contested coastal buffer zone policy has led to conflict and increasing disparities between social groups. The

TABLE 3 Summary of current challenges in EWS implementation organized according to selected resilience elements from the CCR framework and the three dimensions of EWS policy/guidance

Resilience element	Normative challenges	Cognitive challenges	Procedural challenges
Governance: Leadership, legal framework and institutions provide enabling conditions for resilience through community involvement with government	■ EWS development can have bearing on and can be affected by socio-economic factors and trajectories of social change in society and communities	■ Sectoral fragmentation of sub-national DRM planning undermines integrating frameworks and collective actions ■ Lack of mechanisms for facilitating exchange of diverging stakeholder perspectives leads to duplication of efforts and lack of integration	■ Methodological polarization between top-down government approach to provide EW technology and bottom-up NGO approach that focuses on community-based disaster preparedness ■ Lack of mechanism for feeding back CBDRM lessons learnt to the formulation of policy and guidance
Risk knowledge: Leadership and community members are aware of hazards and risk information is utilized when making decisions	■ The norms of stakeholder participation in generating risk knowledge are contested ■ Low trust in EWS providers undermines the confidence in risk knowledge	■ The prevailing political economy of knowledge in EWS development disqualifies competencies of many stakeholders ■ CBDRM guidance/policy promotes knowledge prescriptive and expert-based approaches	■ Risk knowledge for CBDRM is approached as prescriptive awareness programmes ■ EWSs introduce artefacts as systems of symbols that have to be internalized and/or constructed by users ■ Norms of stakeholder participation conflict with the knowledge prescriptive education programme
Warning and evacuation: The community is capable of receiving notifications and alerts of coastal hazards, warning at-risk populations and individuals acting on the alert	■ De facto priorities of national and sub-national decision makers conflict with the stated goals of the regional EWS	■ Decentralization implies additional roles and responsibilities for provincial and district government authorities with limited or no additional financial and human resources ■ Tsunami EWSs dominate and a multi-hazard approach is rarely integrated	■ Despite considerable international funds for the development of national EWSs in the short term, there is a lack of sub-national funds available ■ Reliance on volunteerism and concerns over the long-term sustainability of newly created government institutions, such as research facilities and operational emergency centres

same communities that have been disadvantaged by this policy are now expected to participate in EWS development (Ingram et al., 2006). In Banda Aceh, a similar idea to create a coastal buffer, the so-called 'blue zone', was not enforced because of resistance from fishing communities and the absence of initiatives to support alternative livelihoods. High disaster impacts in Asia are frequently attributed to ineffective measures to address increasing vulnerabilities arising from population growth, poorly planned urbanization and other socio-economic factors. In Krabi Province, Thailand, land acquisition by private investors that are in alliances with influential people in villages and tambons in many cases push poor and marginalized people to areas exposed to hazards.

Due to the power associated with EWSs, that is, in mobilizing large numbers of citizens, the possession and sharing of knowledge can be highly politicized. In Padang, Indonesia, both the provincial and the district government have responsibility for disaster management. Development actors there experience an ongoing tension due to the struggle for authority between the two agencies arising from ambiguities in their respective mandates. In Sri Lanka, staff at the Disaster Management Center (DMC) described similar challenges of power sharing between government departments. In Krabi Province, provincial government officials raised concern that surveys conducted by the Department of Disaster Prevention and Mitigation after the 7 July 2008 drill were not shared with the provincial departments. Moench (2005) argued that the legitimacy of authority can be a dilemma when agencies fight for seizing mandates. Some argue that current policies relating to EWSs support existing power structures but increase social vulnerability when disaster risk management (DRM) is not integrated into wider development policies (Heijmans and Victoria, 2001).

3.1.2. Cognitive: Sectoral fragmentation and duplication of efforts

There is a continuing cognitive challenge in integrating community-based disaster risk management (CBDRM) with sectoral policies, and in developing standard operating procedures (SOPs) at sub-national levels. Linking coastal zone management and disaster preparedness requires the development of new mental models and shared theoretical frameworks for action with new roles and responsibilities, for example, to ensure that resettlement does not occur in areas highly exposed to hazards or in areas important for food production. In many cases, CBDRM practitioners have to frame and repackage EWSs in innovative ways to address other priorities such as livelihoods improvement and overall development planning, particularly when working with large numbers of partners. In Krabi Province, this comprises alignment with restoration and expansion of mangrove ecosystems to increase food security and reduce hazard impacts. In Sri Lanka, there is an ambition to use the 2010 revision of the national coastal zone management plan to enable DRM activities to build on existing management structures and staff resources at the lowest levels of government through integrated coastal zone management and special area management planning.

Divergences in what actors consider desirable benchmarks of institutional coordination and leadership can result in a duplication of efforts and lack of capacities and progress. However, parallel efforts can also be a manifestation of attempts to complement or improve existing initiatives. Some claim that what might be perceived as a duplication of efforts reflects a distribution of labour among government and NGOs (Shaw, 2006). Several NGOs in Sri Lanka regarded their DRM and EWS projects as contributions to government technical instalments, focusing on the strengths of NGOs, namely participatory engagement with communities. One example of this is the Hazinfo project (Evaluating Last Mile Hazard Information Dissemination) led by LIRNE Asia in Sri Lanka (LIRNE Asia, 2008). EWSs tend to evolve in a piecemeal fashion (Davis et al., 1998) and often ride on the back of political and financial windows of opportunity, aligned with past and ongoing initiatives by various stakeholders. For instance, in Sri Lanka, the Disaster Management Act had been in the pipeline long before the 2004 tsunami, but was only officially launched in 2005. Similarly, in Indonesia, the Badan Meteorologi dan Geofisika described how it had attempted to gather support for developing a national EWS in 2003, but only after the 2004 tsunami did local government express full commitment.

3.1.3. Procedural: Polarized approaches and lack of feedback from practice

Normative divergence and cognitive fragmentation can lead to polarizations at the procedural level if mechanisms to create synergies are absent. This is seen most strongly in the tension between what is frequently described as a 'top-down government approach' to provide EW technology ('hardware') and a 'bottom-up NGO approach' that focuses on community-based disaster preparedness ('software'). This is linked to notions of NGOs being 'over-participatory', focusing too strongly on engagement with community organizations that lack legal status, and avoiding the governmental bureaucracy (see also Lukitasari, 2006). In Krabi Province, competition for donor support has undermined previously well-functioning relationships between NGOs. In Indonesia, donor funding has been largely allocated to high-profile programmes, many of which do not pay attention to the actual needs of communities.

Sub-national platforms for dialogue between government, NGOs and CBOs play a crucial role in enabling stakeholders to build operational relationships with communities of practice with other normative and cognitive standpoints. However, despite the growth of sub-national stakeholder networks, there are few adaptive mechanisms that enable feedback of lessons learnt for CBDRM from practice to policy. Many national and international workshops and conferences have been conducted but these events have been criticized for their focus on high-level policy goals, concepts and/or theories. NGOs voiced concerns that government representatives rarely stay on after their own presentations to learn from the ensuing discussions or commit to represent their institutions in stakeholder meetings. Important implementation lessons thus tend to be only communicated to staff working at lower government levels who do not have the mandate to respond to the needs and do not have the capacity or authority to induce positive change within their organizations. Another challenge is the disconnect between expatriate staff, many of whom have limited knowledge of the local context, and national and sub-national stakeholders (see also ALNAP, 2003). Governments are also highly dependent on external consultants because of a lack of capacities to conduct risk assessments. Such concerns motivate criticisms that decisions are made without considering the needs of communities and that the form and procedures in dialogues relating to EWS development should be more critically scrutinized.

3.2. Risk knowledge

3.2.1. Normative: Contested mechanisms for stakeholder participation and distrust in risk information

The challenges associated with developing effective strategies for DRR have in recent years led to an increasing emphasis on participatory planning in humanitarian work with mainstreaming of community participation into international DRR policy and humanitarian standards (ALNAP, 2003; de Ville de Goyet and Morinière, 2006). The Indonesian Disaster Management Law No. 24 of 2007 provides the legal basis for the participation of government, NGOs, the private sector and communities. Many assessments, however, highlight that such participatory approaches to DRR are rarely taken (i.e. IEG, 2006). This is partly due to the fact that goals of community participation in EWS policy reflect contested norms regarding what constitutes good governance (see also Tingsanchali, 2005). Many researchers in the region remain unconvinced of the value of participatory methods, and DRR professionals are reluctant to give up their integrity as 'experts' when the norms of 'people-centred' early warning have not yet been translated into institutional changes in research, agency and government organizations. While one Sri Lankan NGO observed that 'there is a huge gap between the military man and the community', other development professionals are of the opinion that military-controlled approaches to DRR are more efficient than participatory models.

Because they are excluded from national EWS planning decisions and processes, some individuals and communities disregard official warnings and instead take their own initiatives to warn and evacuate. These actions are guided by traditional knowledge of the hydro-meteorological changes (e.g. in the Indonesian Simeulue Islands) and employ TV, walkie-talkies and alternative speaker systems (see also TAW, 2007).

In some villages a lack of trust in the government is rooted in a general suspicion of government agencies because of rumours/evidence of fraud and corruption in post-tsunami compensation payments for lost and damaged items such as boats and fishing gear. This is further exacerbated where resource conflicts have created factions among local user groups. Such tensions are frequently not considered in the negotiation between international donor organizations and national and sub-national decision makers regarding ownership, roles and responsibilities between governments in the region, international donor agencies and UN agencies.

3.2.2. Cognitive: A prescriptive political economy of knowledge

Policy and guidance widely recognize that EWSs require a well-functioning communication system between the organizations comprising the warning chain. Three sets of actors are typically discerned: originators, intermediaries and disseminators or recipients of warning messages (Davis et al., 1998). The ICG/IOTWS Concept of Operations states that in end-to-end early warning 'information must flow from one end (detection) to the other (community response) without interruption or ambiguity (Elliot, 2006, p. 5). The 'last mile' is here situated in the operational context of the recipients, where effective communication depends on the transfer of information from the national warning centre to the communities. This view on communication is rooted in a certain communication model, which by Lackoff and Johnsson (1980) has been termed the 'conduit' metaphor, that is, the assumption that communication comprises the 'conduit', or transfer, of objective information between two or more stakeholders.

When the EWS is designed by the same actors that are tasked with managing it in the warning situation, this communication model is extrapolated from the warning situation to the development of the EWS. The distinction between 'people with information' and 'people at risk' (US-IOTWS, 2007) creates a political economy of knowledge, which disqualifies the competencies of some stakeholders and promotes knowledge prescriptive and expert-based approaches. For instance, the distinction between 'risk management' and 'community participation' in many planning models might disconnect stakeholder involvement from knowledge generation. The attention paid to 'indigenous knowledge' in Krabi Province and in Indonesia is an example of an attempt to counter marginalization of local knowledge. However, this can be problematic if the validity of claims depends on judgements of what comprises 'intergenerational wisdom' or 'indigenousness'. Such judgements are often made by outsiders, for example, government representatives or development actors (see also Bankoff, 2001).

3.2.3. Procedural: Education and awareness raising

A consequence of prescriptive approaches to risk knowledge is to emphasize a one-way teaching model in public education and awareness programmes. Education and awareness raising is deeply engrained as a core element of most EWS frameworks (e.g. Perera, undated; CTEC, 2007). UN/ISDR states that a unanimous view has emerged among humanitarian agencies 'that a better education and awareness of local communities on natural hazards could have contributed to reduce and mitigate the tragic human losses caused by these events' (UN/ISDR, 2007a, p. 11). Indeed, one of the most commonly encountered problem statements of government and NGO staff across the three case studies was that of 'lacking awareness' among communities.

However, EWS development introduces artefacts as systems of symbols that have to be internalized and/or constructed by users if they are to be meaningful. The prescription of what is considered relevant risk knowledge means that warning a efacts such as sign-posts and risk maps are in many cases not relevant in a particular location or for a particular end user. In Krabi Province, evacuation routes were determined by the national government based on the consultation of topographic maps without visual on-site inspection of the characteristics of the village and the surrounding area. Because of a lack of consultation with communities, suitable buildings that could be used as shelters, such as temples and schools, were not considered. In one coastal community, the villagers did not trust the safety of the designated safe area and therefore evacuated to the neighbouring village instead during drills. In another location, a sign pointing to a safe site had been erected, but the shelter was never constructed due to a lack of funds. Another concern was that evacuation signs were only in English. One village rescue team set up additional signs in Thai that advised people to evacuate to the local school and mosque. Similarly, the Sri Lankan Disaster Management Act was criticized by a number of NGOs for lacking appreciation of how local factors such as population density and cultural factors such as religion might affect the effectiveness of SOPs.

Very few documents aimed at guiding EWS development entail critical attention to the degree and nature of stakeholder involvement in the project cycle. In a DRM project cycle used by some organizations in Krabi Province, the evaluation phase is placed immediately before project closure and basic participatory monitoring and evaluation principles are not incorporated. While the HFA recognizes that indicators for quantifying and measuring progress for DRR must be meaningful and credible to a range of stakeholders, it provides no guidance as to how stakeholder participation might take place in order to develop and agree on context-specific indicators (UN/ISDR, 2008).

3.3. Warning and evacuation

3.3.1. Normative: Reconciling EWSs and DRM with other priorities

In many countries in the Indian Ocean region, national-level disaster preparedness planning has been considerably improved following the 2004 tsunami through new policy frameworks and a restructuring of the roles and responsibilities of different government agencies for DRM and early warning (Tsunami Global Lessons Learned Project, 2009). In Sri Lanka, the Disaster Management Act changed the legal environment for DRM and a restructuring of the government is currently ongoing with more power being transferred to the DMC. But despite the increased importance given to disaster preparedness at the national level, decision makers at provincial, district and village level have to reconcile new demands for disaster preparedness placed on them with a range of other priorities. The value sets, willingness and priorities of local administrators to a large degree determine to what extent EWS and DRM policies and strategies are implemented. This is particularly so in decentralized governance systems such as in Indonesia, where ultimate fiscal and managerial responsibilities lie with the district and provincial governments.

There are different perceptions of the role of DRM in relation to other sectors. For example, in Krabi Province, fisheries and tourism are the two most important economic sectors and the provincial government sees the demonstration of DRM activities as important to instil confidence in the tourism sector. In Indonesia, we came across the opposite; some local government authorities resist the implementation of disaster preparedness activities driven by the national government because of concerns of the negative image this might shed on tourism destinations. Similarly, in the communities, the value of CBDRM as a tool for community empowerment depends on the priorities and commitment of local stakeholders to engage in risk reduction activities. Both vary markedly between communities, even in those that were severely affected by the 2004 tsunami. In Krabi Province, the DRM Committee of Ban Tha Klong village developed an EWS for sea-based transport, which contributes to income generation through fishery and trade. Of the 84 households, 60 depend directly on fishery and most others are connected by family ties.

3.3.2 Cognitive: Taking a multi-hazard perspective

There is consensus among policy makers, practitioners and researchers that EWSs should address multiple hazards because such systems enable the integration of hazard information sharing in the context of broader societal development by linking preparedness and response to different kinds of hazards, including hydro-meteorological hazards, as well as for instance biological hazards and hazards to food security (see also Minamiguchi, 2005). The Hyogo Declaration stresses the importance of '[...] integrated, multi-hazard, and multi-sectoral approaches [...]' to build resilience to disasters (UN/ISDR, 2005, p. 2) and the G8 Response to the Indian Ocean Disaster pledged that 'early warning systems should cover as many hazards as possible, not just tsunamis [...]' (G8, 2005).

However, there is little evidence in the countries investigated that national EWSs currently being developed consider any other hazards than tsunamis. The US-IOTWS, which includes Thailand and Sri Lanka, aimed to apply a multi-hazard approach that simultaneously addresses tsunamis as well as other coastal hazards such as cyclones, sea swells, floods and earthquakes (www.iotws.org). However, in the transition workshop the lack of a multi-hazard focus was identified as a major gap (US-IOTWS, 2008). The Indonesian Tsunami Warning System, currently being developed through the German–Indonesian Cooperation for a Tsunami Early Warning System (GITEWS), focuses on tsunamis, earthquakes and volcanic hazards (www.gitews.org). Some representatives from government agencies informed us that they aim to build a multi-hazard system in the longer term, but other stakeholders voiced doubt as to whether this will be accomplished.

Because of the diversity of needs and priorities at the local level, it is important to frame proposed interventions for early warning and DRM in contexts that are relevant in addressing local priorities and building partnerships between different actors. Addressing multiple priorities within an integrated framework will also improve the coordination between different actors. This is particularly so when EWS development as a political and social activity has to be bought into by government representatives as well as community leaders. The Thai Red Cross and the RTF both learnt that DRM activities must be planned with local decision makers such as the village headman to ensure that response plans and EWS activities are properly sanctioned to get buy-in from villagers.

While there is considerable potential for integrating DRM in the environmental domain, there are to date few efforts to integrate DRM into development planning processes in order to reduce disaster risks in the recovery process and to improve livelihoods. The World Conservation Unit in Sri Lanka highlighted the potential to integrate DRM objectives when performing environmental impact assessments. Some actors have started to link DRM with natural resource management. This provides an incentive for communities to engage in DRM and ensures local ownership (see also Sudmeier-Rieux et al., 2006).

3.3.3. Procedural: Creating sustainable mechanisms for DRM funding

Despite considerable international funds for the development of national EWSs in the region, there are many concerns regarding the distribution of funding for early warning and disaster preparedness activities. While many sub-national authorities and non-government actors experience a shortage of funds, strict rules about the use of the significant volume of donations from the international community mean that a large proportion remains unspent almost 5 years after the tsunami. Some organizations lack the capacity to absorb the large amount of funding they received through donations. Also, some donors are not aware of the budgeting procedures in local government agencies (Lukitasari, 2006; ADPC, 2006) and some local government officials are reluctant to provide funds dedicated to CBDRM (Managbanag, 2006).

As a consequence of the lack of resources at the local level, CBDRM relies heavily on volunteerism. The Sri Lankan Red Cross has 100,000 trained volunteers; many of these are pupils recruited in schools. In some villages in Krabi Province, only those who have acted as volunteers are eligible for village elections. However, despite incentives such as health care, skills training and improved social status, and the dedication of volunteers, many stakeholders voiced concerns about the longer-term sustainability of disaster preparedness efforts because even volunteers require basic financial support for operational logistics such as transport, food and compensation for the loss of income. In all three countries, the lack of funds to pay or at least compensate volunteers for their time is a cause of low staff retention rates. The high turnover of volunteers and the need to continuously recruit and train new people throughout project implementation puts a considerable strain on an organization's capacities. Frustration over the lack of resources also relates to the inability of committees and volunteers to act and to induce positive change in their communities. Because many of the volunteers have full-time occupations and no compensation is provided for their time, they may not participate in important meetings. In the Philippines, limited local level funds for the implementation of national disaster preparedness policies means that village Disaster Coordinating Councils/Committees are essentially 'shell' organizations that are not functional in emergency response (Heijmans and Victoria, 2001).

Innovative ways of addressing DRM financing have recently emerged in the form of micro-credit arrangements. In Krabi Province, the establishment of revolving loan funds provides a major entry point for the RTF into EWS and DRM projects with community groups. In Sri Lanka, many community groups that are engaged in NGO and government programmes were originally formed to administer economic activities such as micro-credit schemes or loan funds. In Vietnam, Development Workshop France provides short-term affordable loans to reduce household vulnerability by strengthening house construction (Chantry, 2006). Improved financing mechanisms are important in making DRR more effective (e.g. Southasiadisasters.net, 2005). In Sri Lanka, LIRNE Asia and Sarvodaya developed a new funding model to ensure the retention of staff. In the IOTWS, such community-based micro-financing mechanisms have not received much attention and many of the NGOs supporting such activities play only a minor role in EWS development.

4. Conclusions

The insights obtained from this multi-stakeholder participatory assessment suggest that the critical perspective on the 'last mile' espoused in the HFA and other EWS policy and deliberations is not sufficiently translated into action in the policy implementation process. This represents a major barrier to the development of effective EWSs in the region. There is hence an urgent need to ensure that the emphasis on people-centred warning systems introduced with the HFA does not remain semantics, as is currently perceived by many practitioners.

The evidence presented in this paper illustrates that practitioners operate in contexts shaped by multiple stakeholder agendas and navigate diverse needs and priorities. However, policy recommendations and guidance tend to focus predominantly on the procedural dimension of EWS development. More attention needs to be paid to the cognitive and normative challenges in positioning EWSs in the wider trajectories of social change in societies and communities at risk. Better platforms for knowledge sharing need to be established to enable stakeholders to collectively negotiate these challenges, to improve the integration of early warning with other priorities such as livelihoods improvement, natural resource management and community development, and to provide opportunities for critical reflection of 'on-the-ground' experiences and lessons learnt. Such mechanisms will require that local actors, such as district and municipal government representatives, community leaders and NGO field staff, become more strongly involved in the improvement of national policy objectives and action plans, for instance addressing the need for more sustainable funding mechanisms. However, the normative nature of stakeholder and community participation must be better integrated with existing governance cultures.

We have presented an application of the CCR framework to better appreciate the complexity of the policy implementation process. The original framework does not visually indicate the cross-cutting nature of governance and risk knowledge, but the above analysis highlights cross-cutting issues through the dynamic feedback from and to other elements. In addition to the three dimensions considered here, future work could address more specifically the feedbacks between all resilience elements and explore how unexpected events such as large-scale disasters affect the self-organizing capacities of societies at risk.

Acknowledgements

This paper is an outcome of the SEI Programme *Sustainable Recovery and Resilience Building in the Tsunami Affected Region* supported by the Swedish International Development Cooperation Agency. We would like to thank the Asian Disaster Preparedness Center, Raks Thai Foundation and GTZ International Services for facilitating stakeholder consultations in Sri Lanka, Thailand and Indonesia, respectively. In addition, we would like to extend our gratitude to all early warning system stakeholders in the Indian Ocean Region who have communicated with us throughout the last two years and provided valuable perspectives and insights from their efforts in building more disaster-resilient societies.

References

ADPC, 2006. Learning from Community-Based Practices: Strengthening Policy and Partnerships. *Proceedings of the Fourth Disaster Management Practitioners' Workshop for South-East Asia*. Asian Disaster Risk Preparedness Center, Bangkok, Thailand.

ALNAP, 2003. *Participation by Crisis-Affected Populations in Humanitarian Action. A Handbook for Practitioners, Action Learning Network for Accountability and Performance in Humanitarian Action*. Overseas Development Institute, London.

Andersen, P. S., 2007. Last-mile hazard information dissemination: challenges and opportunities for Sri Lanka. *Asian Disaster Management News*, 13(1). 7–8.

Bankoff, G., 2001. Rendering the world unsafe: 'vulnerability' as western discourse. *Disasters*, 25(1). 19–35.

Beck, U., 1992. *Risk Society: Towards a New Modernity*. Sage, London.

Buchanan-Smith, M. and Davies, S., 1995. *Famine Early Warning and Response – The Missing Link*. Intermediate Technology, London.

Chantry, G., 2006. Microfinance and disaster preparedness: an innovative approach from Vietnam. Financing disaster preparedness at the household level. Partnerships for Disaster Reduction–Southeast Asia (PDRSEA), *PDRSEA Newsletter*, 3(March). 2–3.

CTEC, 2007. *Upgrading a Community-based Public Address System and its Use for Awareness Raising through Child/youth Study Groups (COMPASS)*. Community Tsunami Early-warning Centre, ProVention Research & Action Grants Proposal.

Davis, I., Sanderson, D., Parker, D. and Stack, J., 1998. *The Dissemination of Warning*. UK National Coordination Committee for the IDNDR, Thomas Telford Publishing, London.

De Marchi, B., 2007. Not just a matter of knowledge. The Katrina debacle. *Environmental Hazards*, 7(2). 141–149.

de Ville de Goyet, C. and Morinière, L. C., 2006. *The Role of Needs Assessment in the Tsunami Response*. Tsunami Evaluation Coalition, London.

Douthwaite, B. and Ashby, J., 2005. *Innovation Histories: A Method for Learning from Experience*. ILAC Brief 5, IPGRI, Rome, www.cgiar-ilac.org/downloads/Briefs/Brief5Proof2.pdf.

Elliot, T., 2006. The Indian Ocean tsunami warning and mitigation system: progress and challenges. *Disaster Management News*, 12(4). 4–5.

G8, 2005. *Response to the Indian Ocean Disaster, and Future Action on Disaster Risk Reduction*. G8 Summit, Gleneagles, Scotland, 6–8 July 2005, www.unisdr.org/eng/media-room/press-release/2005/Post-G8_Gleneagles_Tsunami.pdf.

Hamza, M., 2006. *The Human Dimension of Early Warning – Why Technology Alone is Not the Answer, Sustainable Recovery and Resilience Building in the Tsunami Affected Region*. Stockholm Environment Institute (SEI), February (unpublished).

Heijmans, A. and Victoria, L., 2001. *Citizenry-Based and Development-Oriented Disaster Response. Experiences and Practices in Disaster Management of the Citizens' Disaster Response Network in the Philippines*. Center for Disaster Preparedness.

Hollister, D., 2008. UNDP crisis prevention and recovery unit. Jakarta, Indonesia, personal communication, 24 November.

IEG World Bank Independent Evaluation group (IEG), 2006. *Hazards of Nature, Risks to Development*. An IEG Evaluation of World Bank Assistance to Natural Disasters, World Bank, Washington, DC.

Ingram, J. C., Franco, G., Rumbaitis-del Rio, C. and Khazai, B., 2006. Post-disaster recovery dilemmas: challenges in balancing short-term and long-term needs for vulnerability reduction. *Environmental Science and Policy*, 9. 607–613.

IOC, 2008. *Strengthening National Capacities for Tsunami Early Warning and Response Systems in the Indian Ocean*. Progress Report as of 25 March 2008, Indian Ocean Consortium.

IOC, UN/ISDR and USAID, 2005. *Assessment of capacity building requirements for an effective and durable national tsunami warning and mitigation system in the Indian Ocean*. Consolidated Report for 16 Countries Affected by the 26 December 2004 Tsunami. 8 November.

Kelman, I., 2009. Don't forget the people. *Weather and Society Watch*, 3(2). 7, 14.

Lackoff, G. and Johnsson, M., 1980. *Metaphors We Live By*. University of Chicago Press, London.

Lassa, J. A., 2008. *When Heaven (hardly) Meets the Earth: Towards Convergency in Tsunami Early Warning Systems*. Paper presented at the ISSM Conference, Delft, The Netherlands, 13–15 May.

Lindahl, K. B., 2008. *Frame Analysis, Place Perceptions and the Politics of Natural Resource Management: Exploring a Forest Policy Controversy in Sweden*. Doctoral thesis. 2008:60, Swedish University of Agricultural Sciences (SLU).

LIRNE Asia, 2008. *Regional Dissemination of Findings from the Last-Mile Hazard Information Dissemination Pilot Project*. HazInfo Supplemental Report.

Lukitasari, C., 2006. Strengthening local government system to support community initiatives. *Proceedings of the Fourth Disaster Management Practitioners' Workshop for South-East Asia*, ADPC (Asian Disaster Risk Preparedness Center) (ed.). Learning from Community-Based Practices: Strengthening Policy and Partnerships, Bangkok, Thailand. 48–51.

Managbanag, E., 2006. The contribution of the Philippine National Red Cross in capacity building for disaster management in vulnerable communities in partnership with the local government units in the Philippines. *Proceedings of the Fourth Disaster Management Practitioners' Workshop for South-East Asia*, ADPC (Asian Disaster Risk Preparedness Center) (ed.). Learning from Community-Based Practices: Strengthening Policy and Partnerships, Bangkok, Thailand. 45–47.

Minamiguchi, N., 2005. *The Application of Geospatial and Disaster Information for Food Insecurity and Agricultural Drought Monitoring and Assessment by the FAO GIEWS and Asia FIVIMS*. Paper presented at the Workshop on Reducing Food Insecurity Associated with Natural Disasters in Asia and the Pacific Bangkok, Thailand, 27–28 January.

Moench, M., 2005. *Kobe Report Draft*. Report of Session 2,7, Thematic Cluster 2. People Centred Early Warning Systems, p. 3.

Parsons, W., 1995. *Public Policy: An Introduction to the Theory and Practice of Policy Analysis*. Edward Elgar, Cheltenham.

Perera, S., undated. *Disaster Management Programme*. Sri Lanka Red Cross Society (SLRCS), Power Point presentation.

Powell, N. and Jiggins, J., 2003. Learning from participatory land management. *The International Handbook of Social Impact Assessment*, H. A. Becker and F. Vancley (eds). Conceptual and Methodological Advances, Edward Elgar, Cheltenham, UK. 44–55.

PPEW, 2009. Platform for the Promotion of Early Warning. www.unisdr.org/ppew/ppew-index.htm.

Rego, A. J., 2001. *National Disaster Management Information Systems & Networks: An Asian Overview*. Paper presented at Global Disaster Information Network (GDIN), http://unpan1.un.org/intradoc/groups/public/documents/APCITY/UNPAN009640.pdf.

Richardson, D. and Paisley, L. (eds), 1998. *The First Mile of Connectivity, Rome: Food and Agriculture Organization (FAO).* www.fao.org/sd/cddirect/cdpub/SDREpub.htm.

Salter, J., 1996. Towards a better disaster management methodology. *Australian Journal of Emergency Management*, 10(4). 8–16.

Shaw, R., 2006. Critical issues of community based flood mitigation: examples from Bangladesh and Vietnam. *Journal of Science & Culture*, 72(1–2). 62–71.

Singh Bedi, G., 2006. Strengthening multi-hazard early warning systems – the last mile. *Asian Disaster Management News*, 12(4). 7–8.

SLIM, 2004. *The Role of Conducive Policies in Fostering Social Learning for Integrated Management of Water.* Policy Briefing, EU Fifth Framework Programme Research Project, Social Learning for the Integrated Management and Sustainable Use of Water at Catchment Scale.

Smith, D., 2005. Tsunami: a research perspective. *Geology Today*, 21(2). 64–68.

Southasiadisasters.net, 2005. *Disaster Risk Mitigation: Potential of Micro Finance for Tsunami Recovery.* Special Issue 7.

Sudmeier-Rieux, K., Masundire, H., Rizvi, A. and Rietbergen, S. (eds), 2006. *Ecosystems, Livelihoods and Disasters: An Integrated Approach to Disaster Risk Management.* The World Conservation Union-IUCN, Gland, Switzerland.

TAW, 2007. *The Tsunami Early Warning System in Thailand: A Resource Book with a Synopsis of Comments by Tsunami Impacted Communities 30 Months after the Disaster.* Tsunami Aid Watch (ed.), A Programme of the Heinrich Boell Foundation, Southeast Asia Regional Office.

Thomalla, F. and Schmuck, H., 2004. 'We all knew that a cyclone was coming'. Disaster preparedness and the cyclone of 1999 in Orissa, India. *Disasters*, 28(4). 255–269.

Tingsanchali, T. 2005. Development of water-related disaster management in Thailand. *Scientific Forum on Tsunami, its Impact and Recovery*, AIT (Asian Institute of Technology) (ed.), A Regional Symposium, Bangkok, Thailand, 6–7 June, 33.

Tsunami Global Lessons Learned Project, 2009. *The Tsunami Legacy: Innovation, Breakthroughs, and Change.* Published by the Tsunami Global Lessons Learned Project Steering Committee.

Twigg, J., 2003. The human factor in early warnings: risk perception and appropriate communications. *Early Warning Systems for Natural Disaster Reduction*, J. Zschau and A. N. Kuppers (eds). Springer, Berlin.

UN/ISDR, 2005. *The Hyogo Framework for Action 2005–2015: Building the Resilience of Nations and Communities to Disasters.* United Nations International Strategy for Disaster Reduction.

UN/ISDR, 2006a. *Global Survey of Early Warning Systems. An Assessment of Capacities, Gaps and Opportunities Toward Building a Comprehensive Global Early Warning System For All Natural Hazards.* a report prepared at the request of the Secretary-General of the United Nations, United Nations International Strategy for Disaster Reduction.

UN/ISDR, 2006b. *EWC III: Third International Conference on Early Warning, From Concept to Action.* International Strategy for Disaster Reduction, Bonn, Germany, 27–29 March.

UN/ISDR, 2007a. *Disaster Reduction in Asia Pacific.* Issue 3, United Nations International Strategy for Disaster Reduction.

UN/ISDR, 2007b. *Building Disaster Resilient Communities: Good Practices and Lessons Learned.* A Publication of the 'Global Network of NGOs' for Disaster Risk Reduction, International Strategy for Disaster Reduction.

UN/ISDR, 2008. *Indicators of Progress: Guidance on Measuring the Reduction of Disaster Risks and Implementation of the Hyogo Framework for Action.* United Nations International Strategy for Disaster Reduction, Geneva, Switzerland.

US-IOTWS, 2007. *How Resilient is Your Coastal Community? A Guide for Evaluating Coastal Community Resilience to Tsunamis and Other Coastal Hazards.* US Indian Ocean Tsunami Warning System Program supported by the United States Agency for International Development and partners, Bangkok, Thailand.

US-IOTWS, 2008. *Proceedings: Workshop on the Transition of the US IOTWS Program to Indian Ocean Partners.* Bangkok, Thailand, 6–7 December 2007, US Indian Ocean Tsunami Warning System Program.

Wenger, E., 1998. *Communities of Practice: Learning, Meaning, and Identity.* Cambridge University Press, Cambridge.

research paper

ENVIRONMENTAL HAZARDS
Human and Policy Dimensions

Exploring system interactions for building resilience within coastal environments and communities

LORAINE MCFADDEN*

Flood Hazard Research Centre, Middlesex University, London, UK

This paper focuses on identifying examples of first-order system interactions, which make important contributions to building coastal resilience for coastal zone management. This discussion is based on an application of the Institutional Analysis and Development Framework to a case-study analysis of coastal management in South Australia. The study suggests that cross-scale interactions and informal relationships within and between users and managers are key interactions defining resilience outcomes within the current system. A significant constraint on improving resilience was the lack of evaluative criteria for identifying sustainable forms of system behaviour. The paper argues that resilience is a function of a normative statement on the characteristics desirable in the functioning system. Analyses of coastal resilience, which facilitate greater understanding of the range of complexities in coastal behaviour, are therefore central to gaining the most useful insights into the options and pathways for building more sustainable coastal futures.

Keywords: Australia; coastal; knowledge; resilience; scale; systems

1. Introduction: Understanding the complexity of the coastal system for coastal management

A primary emerging message from integrated or adaptive system research is that approaches to knowledge – how it is defined, gathered and applied to problem solving – must imitate to a large degree the patterns and dynamics that define behaviour in the natural environment. This call towards social learning, considering different forms of knowledge as open, interlinked systems in constant dynamics, is currently considered vital towards more meaningful integration (Berkes et al., 2003; Folke, 2006). The move towards exploring interlinked knowledge is not also a call for simple solutions or answers. Rather, it highlights the necessity for environmental management to recognize and understand the complexity of environmental problems – the fact that they are system problems (Holling et al., 1998; Ostrom, 2007). Understanding the patterns of interconnections within systems, and managing these to build more effective network links, is central to the vision for integrated management. This paper focuses on contributing to such system-based knowledge for integrated coastal zone management (ICZM), identifying patterns of network interconnections within coastal systems.

Recognizing and understanding complexity within coastal systems calls us to get better at diagnosing and managing system problems. However, what does doing better mean? Janssen and Ostrom (2006) argue from an institutional perspective that 'getting better' at recognizing complexity involves answering a series of questions relating to the following: (1) social dilemmas, for example, what are the conflicting and comprising interactions between the different agents involved?, (2) uncertainty, for example, how do resource users learn and how do their learning processes differ from how public

■ *Email: L.McFadden@mdx.ac.uk

doi:10.3763/ehaz.2010.0056 ISSN: 1747-7891 (print), 1878-0059 (online) www.earthscan.co.uk/journals/ehaz earthscan

infrastructure providers learn?, and (3) networks, for example, how does information spread among nodes in a network and how does this affect resource management? Applied to research in coastal zone management, some of these dimensions of getting better for ICZM have been explored; for example, what are the barriers and opportunities for increasing learning in developing strategic coastal management strategies (e.g. Bastien-Dalyle et al., 2008; McFadden et al., 2009; Smith et al., 2009)? How do resource users and managers view a functioning coast and what are the obstacles and opportunities for exploring coastal futures (e.g. Milligan et al., 2008; O'Riordan et al., 2008)? There is increasingly widespread awareness of the need to capitalize on shared views and knowledge, building network links in space and time for more sustainable coastal management. However, the fact remains that the specific success factors that can facilitate process patterns of integrated strategic coastal management for the long term are still relatively unknown. With this goal in mind, the primary objective of this paper is to use a case-study discussion of coastal management in South Australia to identify a series of first-order system interconnections that emerge as important to building resilience within the coastal system.

The paper uses a specific institutional framework as a template for identifying and exploring system interconnections, that is, the Institutional Analysis and Development Framework (IADF). This framework emerges from the seminal work of Elinor Ostrom (1990), identifying factors behind institutional choice in natural resources management. It focuses on the interactions among resources (e.g. beach), resource users (e.g. tourists), public infrastructure providers (e.g. government agencies, local users' associations) and public infrastructure (e.g. engineering works, policies and other regulations) to identify configurations that enhance or reduce the robustness of human–natural systems (Anderies et al., 2004) (Figure 1). Robustness is defined on the basis of the maintenance of some desired system characteristics despite fluctuations in the behaviour of its component parts or its environment (Carlson and Doyle, 2002). This means that the concept is clearly situated within a broad resilience framework, with its focus on non-structural dynamical change constrained within a stability domain. The concept of robustness is used within the model to highlight the importance of recognizing both the designed (i.e. robust) and self-organizing (i.e. resilient) components of systems. The key aim of the IADF is diagnostic, identifying why some social–ecological systems are sustainable whereas others collapse. It seeks to achieve this through a framework for organizing relevant variables at a series of different spatial and temporal scales. This paper uses first-order interactions from within the IADF as a template for exploring interactions within the case-study area.

Resilience to natural as well as anthropogenic-induced hazards is a much desired attribute of coastal systems and is widely discussed within coastal management literature. This paper contributes to existing research findings on coastal resilience by providing empirical-based reflections from an institutional perspective on specific system interconnections that may prove particularly effective for restructuring system behaviour towards increased resilience for ICZM. However, this analysis also highlights a contextual challenge in the use of the resilience concept for coastal management. It argues that measures of resilience are often a function of a normative statement on the social, physical or economic conditions considered to be desirable within a particular coastal system. Resilience analyses are therefore vehicles for gaining insights into the different preferred courses of action and the range of choices for managing complex coastal environments. This context of the use of the resilience concept is introduced within the following section. The paper then focuses on the case-study application and discusses a series of emerging themes for improving the broad-scale resiliency of coastal systems. Finally, the case-study analysis is contextualized, revisiting the idea of resilience as a relative concept. This section highlights the importance, to coastal management, of gaining an integrated perspective on the functioning

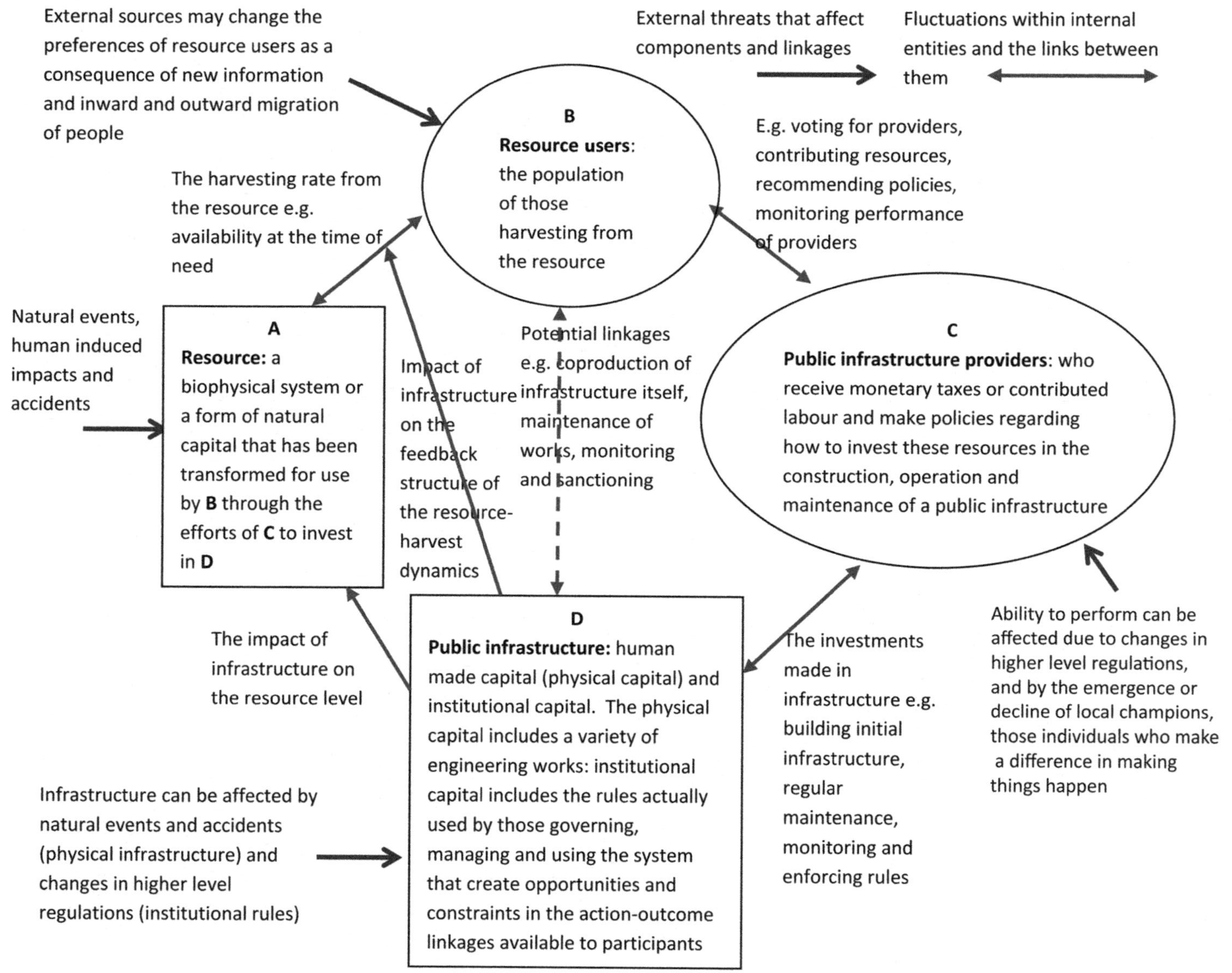

FIGURE 1 A conceptual model of a social-ecological system (Janssen and Ostrom, 2006, after Anderies et al., 2004)

coastal environment and suggests a conceptual framework for coupling models of physical and institutional interconnections for a fuller understanding of the resilience of coastal systems.

2. Characteristics and complexities of resilience for coastal management

The literature specifically focusing on resilience is profuse, opening a field in which a wide range of concepts reside. It has been argued that ideas such as 'resilience', 'robustness' and 'vulnerability' can only be understood in relation to one another (van der Leeuw, 2001). Robustness is mostly used to refer to the structural and other properties of a system that allow it to withstand the influence of disturbances without changing structure or functional dynamics. Current levels of robustness may be based on past adaptations. If these were highly specific, the system may need to adapt upon encountering new types of disturbances (Anderies et al., 2004). As defined by Holling (1973), by contrast, resilience refers to 'the capacity of a system to absorb and utilize or even benefit from perturbations and changes that attain it, and so to persist without a qualitative change in the system's structure'. Such a system may take new external conditions into account by absorbing them into its mode of functioning (Holling, 1986). The difference between the two concepts thus seems to lie in the extent to which (non-structural) changes in dynamics may be introduced into a system under the impact of perturbations. Resilience allows for temporary changes in functioning and dynamics, as long as the system remains within the same stability domain. Finally, vulnerability refers to situations in which neither robustness nor resilience enables a system to survive without structural changes. In such cases, either the system does adapt structurally or it is driven to extinction. All three terms express a temporary condition of the interaction between a system and its context.

The importance of the stability domain to the analysis of resilience highlights the fact that functional systems may be temporally dynamic in form and indeed change may be a form of necessary flexibility to maintain the structure and functional provisions within the system. Within any system there is the possibility of multiple stable states that retain the essential functions of the system. Resilience is therefore a relative concept in the terms that the desirability of any given course of action for increasing resilience depends on the initial starting point within this stability domain. However, the resilience of the system is also relative to the perceived essential functions of the system. There are many actors in the coastal zone who often have different preferences and strongly held beliefs as to the functionality of the coastal system. The stability domain being managed for resilience is defined by choices of actors regarding what are considered to be desirable physical, social and economic attributes of the system. Arguably what is most important for coastal management is mapping these potential stability landscapes, and thus the collision of interests defining the decision-making processes on resilience building, to create the most useful insights into managing the complexities of the system.

This definition of resilience as the capacity of a system to absorb disturbance and re-organize while undergoing change, so as to retain essentially the same function, structure, identity and feedbacks (Walker et al., 2004), has been challenged regarding its adequacy in describing resilient behaviour: particularly in view of the features of complex, adaptive systems. It is argued that resilience is not only about being persistent or robust: it is also about the opportunities that disturbance opens up in terms of recombination of evolved structures and processes, renewal of the system and emergence of new trajectories (Folke, 2006). The future of a system is therefore as much about evolution and emergence into new stability domains, as it is about persistence, and this gives another difficult dimension to understanding future states of the human–physical environment. This evolution of the resilience concept challenges us to consider new, dynamic ways of understanding the behaviour of complex systems.

The current state of art of coastal management is largely focused on a series of different stability domains or landscapes. These range, for example, from the perspective of resource dependencies and the ability of the coastal system to continue to provide a particular range of resources, to the basic continuity of the dynamic physical system as a space to absorb natural and human-induced pressures on the system. At best, we attempt to couple these perspectives for improved ICZM. This paper uses the IADF as a template for the case study and thus focuses on one such landscape, the institutional context as affecting interactions between the physical resource, multiple resource users, and the physical and institutional infrastructure within the system. It seeks to build lessons from this landscape with which to inform integrated assessments of coastal systems. However, the evolutionary perspective of resilience sets a challenging context for coastal management: can our understanding of the resilience of coastal systems for coastal management go beyond simple coupling of physical geomorphological processes (highly dynamic and evolutionary but lacking invention or mutation) with social and ecological responses (that have more complex evolutionary responses with potentials for innovation) to develop newer, dynamic ways of understanding the complex interactions within coastal systems?

3. Key system interactions for building increased resilience: lessons from coastal management in South Australia

The South Australian coastline stretches for more than 4,000 km (including many offshore islands) across commonwealth and state jurisdictions. The natural coastal environment within the region ranges from cliffs, rocky shores and sandy beaches in the South East and West Coast to mudflats, seagrass, samphire and mangrove habitats in the upper St Vincent and Spencer Gulf regions. Approximately 80 per cent of South Australia's population live in the coastal zone, and there is an accelerating trend of population movement to the coast. This 'sea change' is a national phenomenon in Australia, where the rate of population growth in Australian coastal areas is more than 60 per cent higher than the national average (Seachange Taskforce; www.seachangetaskforce.org.au). In the context of South Australia, much of the coastal population is concentrated in metropolitan Adelaide, with the remaining coastal region remote and sparsely settled. Thirty-two coastal councils (in the year 2000) have major statutory functions and responsibilities for local management of the coastal zone, which is both biophysically and socio-economically diverse. There are 24 regional and eight metropolitan councils.

Thom and Harvey (2000) discuss the legacy and contemporary reform of Australian coastal management. Historically, although the Commonwealth (the federal or national government) has previously shown an interest in taking an active role in coastal matters, the management of the coast has been characterized by state domination. Coastal management has continued to be predominantly the responsibility of state and local governments. As discussed by Clarke (2003), the different roles of the three spheres of government, in relation to coastal management, have been pithily described by the following aphorism: 'the Commonwealth has the money; the States the power and Local Government the problems' (Kay and Lester, 1997). However, recent major reviews of Australian national environmental policies and legislation (e.g. Inquiry into climate change and environmental impacts on coastal communities) have signalled a change in the governance framework towards a new model of ICZM that has a stronger national leadership. This proposed Intergovernmental Agreement on the Coastal Zone will be a national cooperative approach and will be overseen by a new Coastal Zone Ministerial Committee. At the state level, two government agencies are currently responsible for managing the South Australian coast (Clarke, 2003). One is concerned with development and has a strong planning focus, whereas the other manages the technical day-to-day maintenance of coastal facilities. The role of the local council in coastal management in South Australia is primarily to implement coastal planning principles and the objectives from the States Development Plan (Harvey et al., 2002).

The following case-study analysis is based on a first-order application of the IADF to coastal management within South Australia, using themes as highlighted by the model as a template for mapping system interconnections. The case-study application is exploratory in nature, based on the analysis of a limited number of in-depth interviews with key actors in local and state government and with coastal academic experts. Data gathering from within the interviews was supplemented by a desktop analysis of reports, journal articles, policy documents and other published material, which provided some overview of the network of interactions within the coastal system. Figures 2–4 outline the overview from this desktop analysis. Figure 2 highlights examples of the impact of public infrastructure on coastal system resources. The focus of Figure 3 is on identifying examples of the interactions between public infrastructure providers (e.g. resource managers) and public infrastructure (e.g. physical and institutional capital). Figure 4 highlights key interactions between resource users and resource managers, those public infrastructures providers who make policies on how to invest in the construction, operation and maintenance of the public infrastructure to manage the coastal zone resources. The figures focus on the links or interactions between the various components of the system, identifying examples of interactions and potential problems for increasing sustainability of the system. Three broad themes with regard to system interactions emerge from the interview and desktop analysis: (1) the role of cross-scale interactions in defining challenges for increasing the resilience of the coastal system, (2) the importance of knowledge exchange and interactions that facilitate learning, particularly between informal and formal system components, and (3) the lack of performance measures as a barrier to building more resilient coastal systems. The themes are discussed in the following sections.

3.1. Role of cross-scale interactions in defining challenges for increasing resilience

Many studies of ICZM have highlighted the necessity for greater clarity in the various responsibilities of local and state government in relation to coastal management. This challenge also emerges from within the South Australian coastal context. The interviews highlighted the clear tension in existing interactions from within the governance components, discussing the difficulties stemming from ineffective interactions between the local, state and federal government. These challenges particularly emerge within an assessment of the linkages between physical and institutional capital and the biophysical state of the coastal resource (Figure 2). The impact of public infrastructure on increasing the resilience of natural resources is seen to be constrained (less effective or ineffective), due particularly to problems in the application of the institutional capital or rules for managing these resources. Figure 2 highlights some of these constraints: for example, inadequate definitions of roles and responsibility between state agencies and between state and local government, as well as between state and federal level; inadequate and overstretched planning controls as reflecting lack of state support at the local council level. Within the system, these interactions in turn emerge from internal conflicts within the public service providers' group of actors (i.e. the resource managers) and the impact of this conflict on investing, maintaining and enforcing the rules (Figure 3). Such conflicts relate, as an example, to perceptions that federal government uses community organization as a 'cost-cutting exercise' replacing, and hence reducing investment in, permanent civil services at the state and local level. Scaling challenges within the institutional context of coastal management emerge as a strong influence constraining the effectiveness of network links within the system.

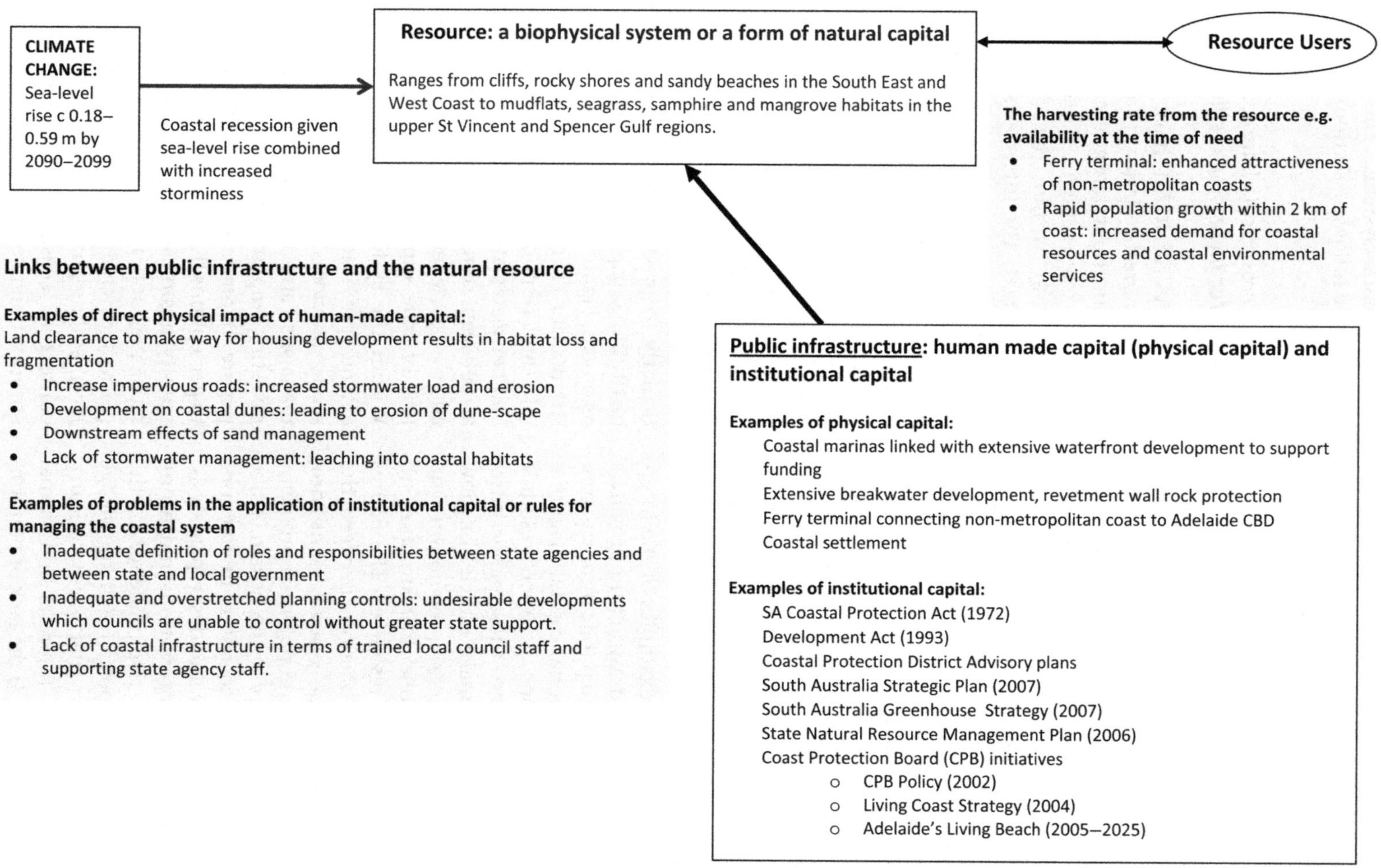

FIGURE 2 Identifying examples of key links between public infrastructure, resource users and the physical resources

<u>Public infrastructure</u>: human made capital (physical capital) and institutional capital (from Figure 2)

Examples of physical capital:

- Coastal marinas linked with extensive waterfront development to support funding
- Extensive breakwater development, revetment wall rock protection
- Ferry terminal connecting non-metropolitan coast to Adelaide CBD
- Coastal settlement

Examples of institutional capital:

- SA Coastal Protection Act (1972)
- Development Act (1993)
- Coastal Protection District Advisory plans
- South Australia Strategic Plan (2007)
- South Australia Greenhouse Strategy (2007)
- State Natural Resource Management Plan (2006)
- Coast Protection Board (CPB) initiatives
  - CPB Policy (2002)
  - Living Coast Strategy (2004)
  - Adelaide's Living Beach (2005–2025)

Public infrastructure providers:

- Local district council are primary decision-making authority
- Australian Local Government Association: common voice – assistance to local councils
- Coastal Protection Board: subject to control of Minister of Environment and Conservation, manages in conjunction with local district councils. Also works in conjunction with regional natural resources management (NRM) boards
- South Australia Government: Coast and Marine Conservation Services, Coastal Management Board and Coast Protection Service
- Natural Resource Management Regions in South Australia

Interactions between the public infrastructure providers (managers) and the public infrastructure i.e. the investments made in infrastructure e.g. building initial infrastructure, regular maintenance, monitoring and enforcing rules

- Commonwealth using community organizations as 'cost-cutting exercise' replacing permanent civil services
- Commonwealth bypassing state government through NRM funding arrangements
- Waterfront development – maintenance breakwater development, revetment wall rock protection and ongoing dredging to protect and maintain an open channel to the sea
- Maintaining and upgrading sea walls, groynes and offshore breakwaters
- Continuing beach replenishment: around 1,500,000m^3 of sand has been brought to Adelaide's beaches mainly by dredging offshore sand sources
- Sea-level rise projections: Metropolitan Adelaide likely to increase the cost of sand replenishment by at least AU$1.25 million per annum
- Rural councils receive funding imbalance due to high spending on metropolitan beach maintenance
- Monitoring programme for environmental change e.g. beach profiling, water quality monitoring: local councils and community dune-care groups
- Australia is one of the major Environmental Impact Assessment jurisdictions in the world: however monitoring and auditing remain the weakest areas, requirements are discretionary

FIGURE 3 Identifying examples of key public infrastructure providers and of linkages between this group of actors and actual investment in public infrastructure

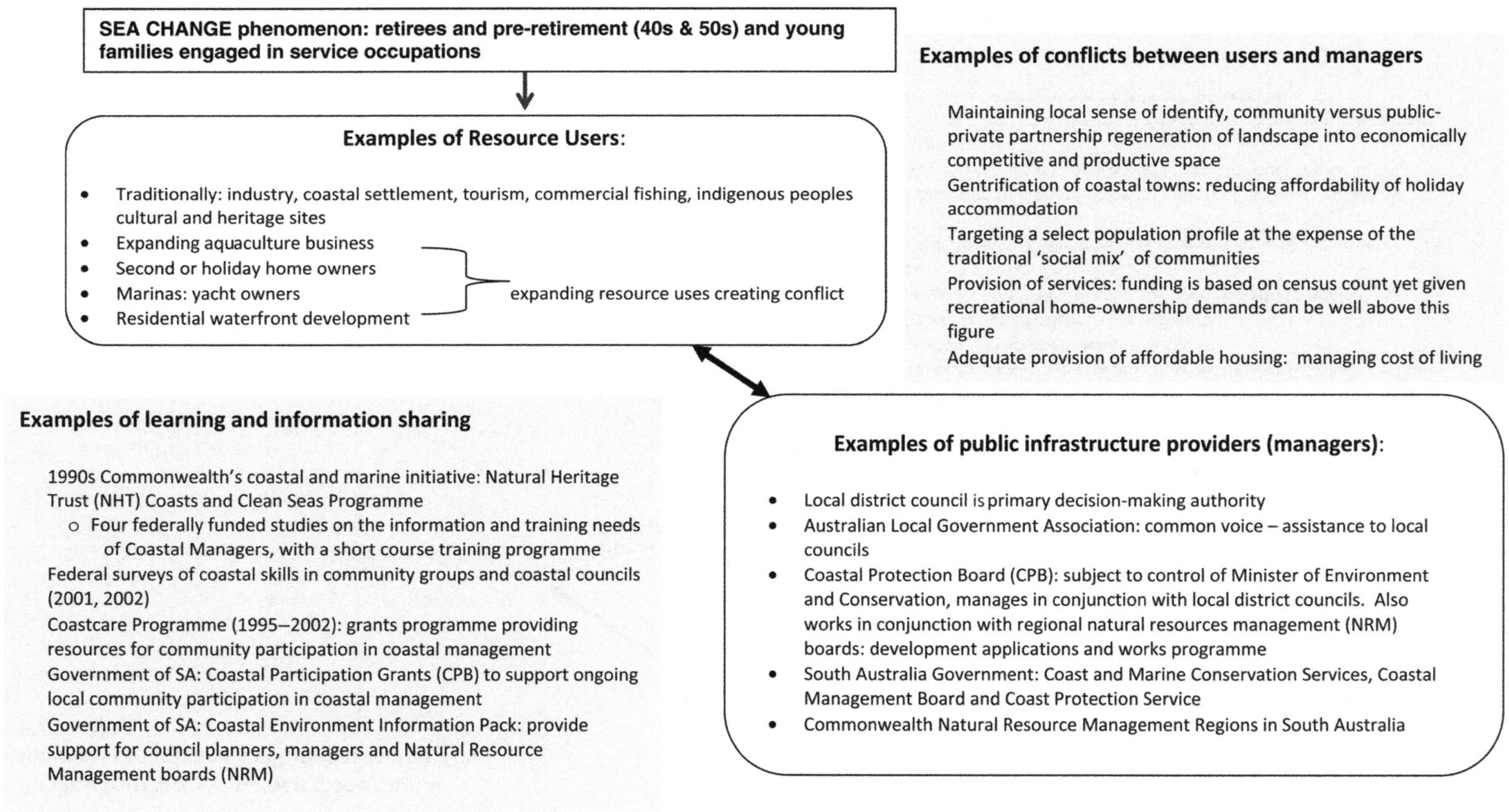

FIGURE 4 Identifying examples of interactions within the case study between public service providers and resource users

The interviews highlighted the fact that these ineffective relations are underpinned to a significant degree by political changes under the current liberal government, which mean that on-ground advice to coastal councils and access to funding come directly from the federal level, largely bypassing state government. This was seen, particularly by the interviewees at the state level, to reduce state instrumentality and the role of the state in coastal management to managing hard protection, although they retain a degree of power regarding development control. A consistent perspective emerging from across the interviewees was that a strong disconnect particularly exists between the state and federal levels and that this has a negative influence on building effective coastal zone management policies and strategies. The tension between state and federal governments also emerged within the interviews in the discussion of networks of information exchange. It related to groups directly lobbying federal government so that state levels have less or limited influence on the decisions relating to these special interest groups. An example given was the National Seachange Taskforce representing coastal councils and local communities, which is a strong and influential group on policy making for coastal management. However, it is a network within which the state government has no direct discussion.

This theme reinforces the centrality of cross-scale interactions to the resilience of coastal systems. One important take-home message from this analysis is that building more effective cross-scale interactions between the various levels of governance within the region has a key role to play in promoting more sustainable approaches to coastal management. Achieving such interactions depends on a greater understanding of structural relationships between the governing scales. An important related point is that redundancy and diversity in institutions play a key role in building resilience, providing 'back-up' options when governance systems fail to respond and allowing for prompt reorganization after a shock (Low et al., 2003). Addressing the cross-scale management challenges reflected in the case study depends on understanding what are the most sustainable patterns of resourcing, operational monitoring and enforcing and policy-making processes, with redundancy and diversity factored into this analysis. This necessitates deliberation between the various scales of governance.

One of the strongest emerging points from the interviews related to asymmetry within the region's network for coastal management. That is, the imbalance that exists between the capacity for coastal management between different local councils. These limits of professional capacity across the councils were viewed at the local scale as the single most important factor in limiting sustainable integrated management of the coast within the state. Given that some local councils within the state are small, they have very limited capacity and might not have a professional planner or professional knowledge of the coast: for example, Kangaroo Island, which has a population of 20,000 people. With a state population of just under 2 million, half of the population reside in the greater Adelaide area. The view from the local level is that the professional interest and expertise is concentrated in Adelaide and particularly within the state government agencies. This presents a significant problem, because the mechanisms are very limited whereby this professional capacity is available to assist local government in the shared task of management. The interviewee highlighted the fact that there are individuals in local government who make a significant contribution to managing the coastal resource system through the expertise that they have accumulated over many years in the field. However, when these 'linchpin people' leave their current positions, this will leave a significant gap in the network, which will have major repercussions for the resilience of the system. An alternative perspective emerged from the interview with state level government. In this case, the accumulated expertise within the Coastal Management Branch of the Department of Environment and Heritage is valued as a centralized professional expert group. A strong feeling from this group is that any attempts to decentralize this expertise will reduce the critical mass of professional experience and have major negative impacts on the provision of coastal management within the region. At the regional level, the critical mass of expertise is seen to produce conditions that are conducive to efficiency: at a local level, interest is more sensitive to principles of equity. The interviews suggest that working through, in very practical terms, the relation between these two principles of sustainability is important to designing more resilient configuration for decision making.

3.2. Importance of knowledge exchange and interactions that faciliate learning, particularly between informal and formal system spaces

A key theme to emerge from within the interviews was the high value placed on the coastal environment within local communities. This was perceived by the interviewees to relate to the ethos that the beach is a 'very public good', that is, it 'belongs to everyone'. The perspective raised was that the Australians put great value on 'Joe Blogs on a modest income being able to access the coast in the same manner as Mr Mansion'. This gives a strong cultural imperative towards maintaining rights and obligations for coastal zone management. This was related by the interviewees to the perceived importance within the Australian culture of the willingness to 'speak out' and 'stand up' and in turn to effective information exchange between local communities and local decision makers.

Further evidence of the importance of informal rules (i.e. cultural, moral or ethical) in defining attitudes and expectations from the coastal system emerged during the interview discussion. In Australia in general and in South Australia in particular, the natural coastal resource system is large, expansive and sparsely settled. Although the expansive nature of the coastal system produces significant challenges for monitoring and enforcement, the size of the resource was perceived to provide a key driving force for incentivizing behaviour towards maintaining and improving the coastal resource. This was related by the interviewees to the strength and attraction of the coast in terms of the value that is placed within Australian culture on space and distance. An additional point raised within this discussion was that often the 'beach is the community' that is the social space where people meet, relax and talk. There is a clear emphasis emerging from the interviews on the importance of culture in characterizing resource users' perspectives on the coastal environment.

This emerging message highlights the strength of the attraction of coasts within Australian culture and, in doing so, points to the importance of effective links between the informal rules that constitute and define these perspectives and the formal policies and regulations governing the coastal environment. The importance of informal space as a vehicle for developing relationships that are meaningful to individuals and groups, enabling flows of knowledge and learning, is highlighted by the literature on social learning (e.g. Pelling et al., 2008). However, the challenge exists of effectively integrating learning from within such informal space with formal rules and regulations that define the behaviours of the organizations or institutions managing the coastal system. Discussions within the interviews suggested that while particular forms of information, for example, user information, project information and news, may be exchanged relatively well, processes of dialogue that facilitate collaborative and deliberative approaches to decision making are less well formulated. The inherent value placed on the physical coastal environment within the Australian social fabric suggests the importance to coastal management of harnessing the networks and experiences of coastal resource users. It also points to the potential usefulness of participatory processes to share knowledge, discuss and deliberate on recommended policies and management actions, with the aim of further increasing incentives within resource user groups towards building resilient coastal systems.

Figure 4 outlines examples of learning and information sharing between resource users and public infrastructure providers. These processes take the form of federal and state-funded initiatives on capacity building for coastal managers and local community participation in coastal management, ranging, for example, from grants for local community participation, surveys of skills and needs of managers, and training to support mangers at local and state scales. The evidence points to a relatively significant legacy of processes for learning and information sharing across the coastal system, as well as ongoing processes of facilitating some exchange of information for coastal management. However, Figure 4 also identifies examples of conflicts or tensions between the two groups of actors. These largely focus around the maintenance of the cultural and social identities of the coastal towns and suburbs against government-driven processes for change of this identity towards economic regeneration. It points again to the central role that the cultural context has in determining the effectiveness of interactions between the two actor groups within policy making relevant to building resilient coastal communities.

One simple perspective that could emerge from this first-order analysis is that although the state has a legacy of government-initiated participation vehicles, a sharp conflict remains between local vision for the local community and the vision that drives public–private partnership pressures on these communities. At one level there would seem to be discontinuity between the learning processes and the impact of these processes on the two actor groups. The ineffectiveness of these learning processes within the region is already reflected in published literature and a number of contributing factors are highlighted, for example, the inability to develop lasting partnerships with the community, the reluctance of government to share power to allow effective participation (Lazarow, 2006) and the absence of review indicators for the success of community participation schemes (Clarke, 2006). The perspective of ineffective learning needs to be underpinned by an understanding of the fact that learning processes are complex and are distributed unevenly among society. They depend on the tasks and roles that actors and organizations play in their contexts of action as well as on the power and abilities they hold (Tàbara et al., 2005). Historical contexts, the continuity of learning and information sharing processes, the degree of participation and power inequalities are also examples of barriers to learning. The analysis suggests that a greater understanding of the constraints and opportunities that define the learning potential, particularly between informal and formal system interactions, may create useful pathways towards building more resilient, sustainable systems.

3.3. A lack of performance measures to assess outcomes of system interactions is a barrier to building resilient coastal systems

A third theme to emerge from the series of interviews relates to scientific knowledge and how this information is transferred across the network of actors in coastal management. It was considered by the interviewees that exchanges of research and scientific data could be significantly improved between the scientific community and policy makers, as well as exchanges within the policy-making community. This was seen as a particularly relevant challenge for management at the level of local government, where ineffective transfer of knowledge to the local level acts as a significant barrier for improving integrated management of coastal resources. A key constraint was also seen to be the lack of mechanisms by which various sources of information are brought together and maintained as a continuous and long-term investment.

This context of the need for continuous and long-term investment in data also emerged when considering the linkages between public infrastructure providers and the range of public infrastructure affecting how the system functions over time (Figure 3). It proved challenging within the context of this exploratory analysis to identify emerging lessons from the interaction between these components for improving the resilience of the system. A critical barrier was the difficulty in obtaining an overview of the effectiveness or sustainability of these various system interactions. There are a range of both national and state-based analyses of coastal vulnerability of the South Australian coastline (Harvey and Woodroffe, 2008). However, performance measures of the behaviour of the coastal system – social outcome measures (e.g. efficiency, equity and accountability) as well as geomorphological and ecological outcomes (e.g. resilience, diversity and redundancy) – were difficult to obtain, and in some instances (e.g. social indicators) do not exist. An analytical review of both the impact of public infrastructure on the feedback structure of the resource and the effectiveness of rules used for governing, managing and using the system becomes difficult in the absence of this information. Thus, the sustainability of the interactions and the possibility for restructuring towards resilience cannot be effectively explored. An important emerging message is that developing such diagnostic skills is central to also developing the capacity to predict, explain and improve the sustainability of the system. This is critical to gaining meaningful insights into how resilient and sustainable the current configuration of users, governance system and resource system is, as well as insights into how this might be further improved.

Another significant barrier to understanding interactions within the case study was the relative lack of analytical material on the interface between governance components and the physical resource. There is a significant body of detailed information discussing elements of the physical resources within the region (e.g. Bowman and Harvey, 1986; Bourman et al., 2000; Harvey et al., 2006). However, there is limited discussion within the literature as to how these resources are linked in behaviour through space and time in system dynamics: the network links within the physical subcomponent of the system. Nor was it easy to obtain information on how such physical system knowledge interfaces with management plans and policies for the region. The evidence indicates that this interaction is likely to be minimal; for example, the Adelaide Living Beach Strategy is built on a series of notional physical resource units that do not relate to the dynamics of the biophysical resource. In practice, this was an area of coastal management strategy development within the state that needed to be considerably strengthened. It is a critically important aspect of building resilience because it relates directly to the centrality of understanding what makes the South Australian coastal zones viable as a functioning resource system.

4. Contextualizing the perspective of resilience emerging from the exploratory analysis

As highlighted in a previous section, by using the IADF as a template for this case study, this analysis of resilience is embedded in one stability landscape that is focused on the sustainability of rules defining the governance, management and use of coastal resources: that is, the institutional context of coastal management as affecting the sustainability of the system through interactions between the physical resource, multiple resource users, and the physical and institutional public infrastructure. There is significant theoretical value to building an assessment of coastal systems for coastal management on a governance-based model. Governance is about power, who has it, who should have it, how it should be exercised and for what purposes. It is about how resources should be allocated, what are the goals of society and how society should be organized (Green, 2009). Thus, governance is one of the fundamental essences of ICZM. The fragmented nature of institutional frameworks and the challenges of adopting stakeholder-based decision making mean that improving governance is central to ICZM. This provides a strong context for the usefulness to coastal management of adopting an institutional-based approach such as that represented by the IADF.

A governance-based model commonly referred to within ICZM literature is the Four Orders of Outcomes by Olsen (2003). This model aims to group together the sequences of institutional, behavioural and social/environmental changes that can lead to more sustainable forms of coastal development. The Four Orders of Outcomes are goals or indicators of sustained positive progress towards ICZM. The IADF could bring significant added value to this approach for understanding governance dimensions of coastal system behaviour. Focusing on diagnosing the causal patterns of interactions within the coastal system that affect ICZM outcomes, the IADF brings a critical depth of analysis to an outcome-based approach. The framework of institutional attributes and variables within IADF provides a comprehensive foundation for thinking about which aspects of the system are likely to have a major impact on the patterns of interactions and the outcomes of these interactions. A key strength of the framework is therefore the opportunity it affords for a detailed and systematic understanding of the governance components of coastal systems. Increasing our understanding of the roles of organizations and institutions within decision-making processes and the nature of rules defining the governance, management and use of coastal resources could considerably strengthen an analytical approach for ICZM.

However, as an institutional model, the IADF approach has little facility for identifying the essence of what makes the coastal system a viable biophysical system. Physical coastal vulnerability is often associated with coastal sensitivity, that is, the relationship of disturbance event frequency to relaxation time of the physical environment (the time taken for the coastal feature to recover its form) (Pethick and Crooks, 2000). Anthropogenic occupation and functionality of coastal zones means that such a physical system orientation of resilience is ineffective as a stand-alone model for coastal management (Orford et al., 2006). However, the perspective captured within this approach reflects an alternative and viable stability domain to that within the institutional model. From this perspective of coastal sensitivity, the resilience focus is on management for change within the physical environment, rather than managing the system to maintain its capacity to provide resources or services to society.

These two perspectives on resilience of the coastal system can potentially be coupled. For example, the analysis of resource unit dependencies within IADF could have a more explicit link with neighbouring resource units at the same scale within the physical behavioural system. A basic parameter of geomorphic systems analysis is that the evolution of one particular element of the coast is strongly influenced by, and influential on, evolution in adjacent areas (Burgess et al., 2002). At a first-order level, this could be achieved by a simple conceptual development of the IADF, highlighting the significance of interactions between adjacent coastal resource units to the assessment of the coastal system resilience. The depth and complexity of this analysis could be further increased within a nested, multi-scale analysis of the physical behaviour systems (e.g. Halcrow, 2002). Exploring the range of interactions with the governance components of the system, across such physical behavioural scales of analysis, provides a fuller perspective of the complexities of coastal change. Such a physical behavioural analysis could itself be embedded within a Source Pathway Receptor (SPR) model, which highlights the pathways within the physical and human environment through which hazards (sources) are transferred to communities, society and the natural environment (receptors) (DETR et al., 2000) (Figure 5). This allows a valuable added perspective in that the geomorphic system is seen itself as an entity that may be harmed (i.e. it is a receptor as well as a pathway) irrespective of human presence (Evans et al., 2004). The combined physical behavioural and SPR analysis facilitates a comprehensive understanding of the nature of the physical environment as a viable system of systems and its consequent role in structuring and defining interconnections within the complex functioning coastal environment. Coupling this analysis with the IADF provides a strong conceptual framework for exploring the potentials in ICZM for managing processes of change within the coastal systems towards more resilient,

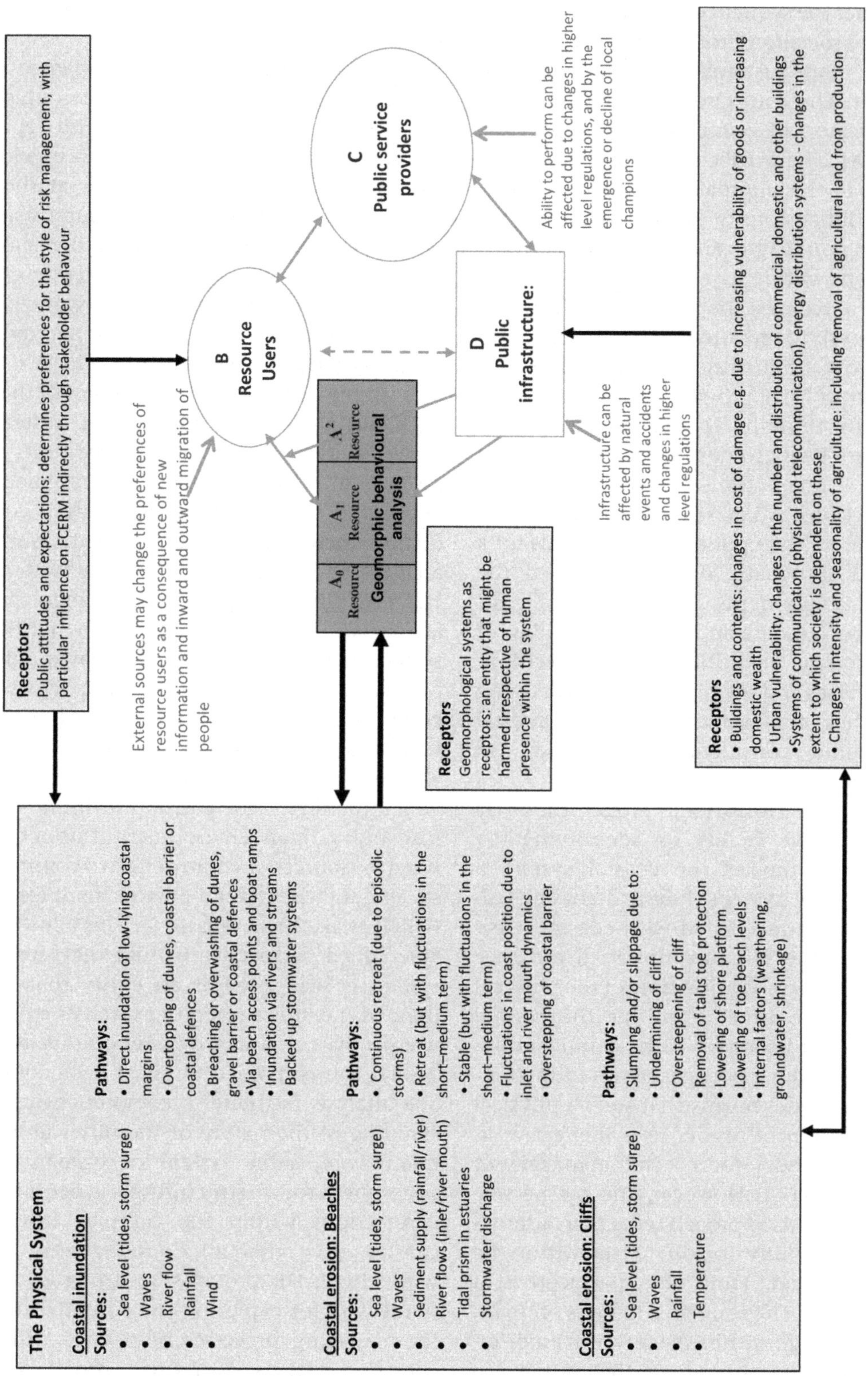

FIGURE 5 A coupled model for understanding coastal system interactions: embedding physical behavioural analysis within the Source Pathway Model and linking this with the IADF approach

sustainable outcomes. Such a proposed conceptual framework is, in essence, a simple development, focused on meshing existing models for exploring coastal systems: the challenge of going beyond this approach to developing transdisciplinary models for understanding resilience remains significant within ICZM. However, reflections on the case-study analysis would suggest that there are still important lessons to be learned for coastal management from such a coupled approach to integrated assessment.

5. Conclusion

Exploring ideas from the IADF within the South Australian coastal management context has identified a range of interactions as potentially significant drivers for increasing the resilience of the coastal environment. These interactions may be summarized under three broad themes. The first widely recurring theme is the importance of scaling issues in determining resilience and the need to better understand cross-scale interactions to build capacity for increasing resilience within coastal systems. This necessitates increasing our understanding of what are the most sustainable patterns of resourcing, operational monitoring and enforcing coastal management, and includes exploring the relations between, and appropriate balance of, equity and efficiency in cross-scale governance processes. The second theme relates to the prominent role that informal rules and interactions play in constructing the behaviour of resource users and managers of the coastal system and highlights the need to integrate these informal components into the rules and policies driving coastal management. Culture and societal norms proved to be a very important influence in the case study in relation to attitudes to the coast and expectations for coastal management, as well as on the effectiveness of network links between users and managers of the system. Increasing existing understanding of the complexities of learning processes is important to facilitating greater exchange between the formal and informal spaces within the system. A third emerging theme relates to the lack of performance measures for assessing the outcomes of interactions within the system and highlights this issue as a significant barrier to building more resilient coastal environments. Developing diagnostic skills and identifying social as well as physical outcome measures is important to gaining meaningful insights into how sustainable the current system configuration is as well as insights into opportunities for enhancing coastal resilience. The primary conclusion from this exploratory analysis is that broad-scale resilience of the coastal system could be substantially improved by focusing attention on a better understanding of, and facilitation between, informal and formal knowledge and processes of decision making and on the scaling challenges within the region, supporting these with a careful, well-constructed analysis of the evaluative criteria of physical, social and institutional systems.

However, analyses of the resilience of coastal systems are often a function of a normative statement as to what is perceived as a desirable stability domain for the system. The lessons identified above, for example, are embedded in a perspective of exploring the sustainability of rules defining the governance, management and use of coastal resources. They emerge from the institutional context of coastal management as affecting the sustainability of the system through interactions between the physical resource, multiple resource users, and physical and institutional public infrastructure. A comprehensive understanding of the complex interactions of the coastal system requires coupling of the institutional perspective on resilience with bio-geomorphological analyses of the processes defining a resilient coastal environment. This paper has suggested, in essence, a simple development of the IADF approach to allow a fuller understanding of the role of the geomorphic component in the functioning coastal system. Integrated analyses are critical to gaining the most useful insights into pathways for building more resilient coastal futures.

Increasing our understanding of opportunities and barriers for improving coastal resilience ultimately depends on building lessons from the application of resilience ideas to a wide range of case studies of differing physical, social and ecological contexts and legacies. These case studies need to be conducted within a theoretical framework that recognizes the complexity defining how combinations of variables affect the behaviour of the coastal system. This paper has presented one exploratory analysis of South Australian coastal management focused on an assessment of resilience based on an institutional analysis. There is a clear need to continue to build comprehensive case-study examples from which comparisons can be used to more confidently identify the strategic system variables and relationships that can most significantly contribute to the resilience of coastal systems.

Acknowledgements

My particular thanks extend to Nick Harvey for hosting me at the University of Adelaide and facilitating a thoroughly interesting and enjoyable research visit. I am also very grateful to Brian Caton and Beverly Clarke for interesting and helpful discussions. Any errors in facts related to the case study or misinterpretation of these facts within the paper are solely attributable to me. Audun Sandberg provided helpful comments on an earlier version of this paper. I also acknowledge particularly useful reviews of this paper by two unknown reviewers.

References

Anderies, J. M., Janssen, M. A. and Ostrom, E., 2004. A framework to analyze the robustness of social-ecological systems from an institutional perspective. *Ecology and Society*, 9(1). 18.

Bastien-Dalyle, S., Vanderlinden, J.-P. and Chouinard, O., 2008. Learning the ropes: lessons in integrated management of coastal resources in Canada's maritime provinces. *Ocean and Coastal Management*, 51. 96–125.

Berkes, F., Colding, J. and Folke, C., 2003. *Navigating Social-Ecological Systems*. Cambridge University Press, New York.

Bourman, R. P., Belperio, A. P., Murray-Wallace, C. and Harvey, N., 2000. Rapid coastal geomorphic change in the river Murray Estuary of Australia. *Marine Geology*, 170. 141–168.

Bowman, G. and Harvey, N., 1986. Geomorphic evolution of a Holocene beach-ridge complex, LeFerve Peninsula, South Australia. *Journal of Coastal Research*, 2(3). 345–362.

Burgess, K., Jay, H. and Hosking, A., 2002. FUTURE-COAST: Predicting the Future Coastal Evolution of England and Wales. *Littoral 2002: The Changing Coast*, EUROCOAST/EUCC, Porto, Portugal. 295–301.

Carlson, J. M. and Doyle, J., 2002. Complexity and robustness. Proceedings of the National Academy of Science, 99(Suppl. 1). 2538–2545.

Clarke, B., 2003. Coastcare, Australia's community-based coastal management program: an effective model of integrated coastal management. Unpublished Ph.D. thesis, The University of Adelaide.

Clarke, B., 2006. Australia's coastcare program (1995–2002): its purpose, components and outcomes. *Geographical Research*, 44(3). 310–322.

Department of the Environment, Transport and Regions, Environment Agency, Institute for Environment and Health, 2000. *Guidelines for Environmental Risk Assessment and Management*. The Stationary Office, London.

Evans, E., Ashley, R., Hall, J., Penning-Rowsell, E., Saul, A., Sayer, P., Thorne, C. and Watkinson, A., 2004. *Foresight. Future Flooding. Scientific Summary Volume 1 – Future Risks and their Drivers*. Office of Science and Technology, London, UK.

Folke, C., 2006. Resilience: the emergence of a perspective for social-ecological systems analysis. *Global Environmental Change*, 16. 253–267.

Green, C., 2009. *Procedural Equality*. SWITCH Project Report. www.switchurbanwater.eu.

Halcrow, 2002. *Futurecoast* (CD-ROM). Produced for Department for Environment, Food and Rural Affairs.

Harvey, N. and Woodroffe, C. D., 2008. Australian approaches to coastal vulnerability assessment. *Sustainability Science*, 3(1). 67–87.

Harvey, N., Bourman, R. P. and James, K., 2006. Evolution of the Younghusband Peninsula, South Australia: new evidence from the northern tip. *South Australian Geographical Journal*, 105. 37–50.

Harvey, N., Clarke, B. and Baumgarten, P., 2002. Coastal management training needs in Australia. *Ocean and Coastal Management*, 45(1). 1–18.

Holling, C. M., 1986. The resilience of terrestrial ecosystems: local surprise and global change. *Sustainable Development of the Biosphere*, W. C. Clark and R. E.

Munn (eds). Cambridge University Press, Cambridge. 292–317.

Holling, C. S., 1973. Resilience and stability of ecological systems. *Annual Review of Ecology and Systematics*, 4. 1–23.

Holling, C. S., Berkes, F. and Folke, C., 1998. Linking Social and Ecological Systems. *Sustainable Development of the Biosphere*, F. Berkes and C. Folke (eds). Cambridge University Press, Cambridge, UK, 342–362.

Janssen, M. A. and Ostrom, E., 2006. Governing social-ecological systems. Chapter 30. *Handbook of Computational Economics*, Vol. 2, L. Tesfatsion and K. L. Judd (eds). Elsevier, Oxford.

Kay, R. and Lester, C., 1997. Benchmarking the future direction for coastal management in Australia. *Coastal Management*, 25. 265–292.

Lazarow, N., 2006. *Community Participation in ICZM: Lessons and Areas for Improvement in Governance. Coastal Management in Australia: Key Institutional and Governance Issues for Coastal Natural Resource Management and Planning*. Cooperative Research Centre for Coastal Zone, Estuary and Waterway Management.

Low, B., Ostrom, E., Simon, C. and Wilson, J., 2003. Redundancy and diversity: do they influence optimal management? *Navigating Social-Ecological Systems: Building Resilience for Complexity and Change*, F. Berkes, J. Colding and C. Folke (eds). Cambridge University Press, New York. 83–114.

McFadden, L., Tapsell, S. and Penning-Rowsell, E., 2009. Strategic coastal flood risk management in practice: actors' perspectives on the integration in flood risk management process in London and the Thames Estuary. *Journal of Ocean and Coastal Management*, 52(12). 636–645.

Milligan, J., O'Riordan, T., Nicholson-Cole, S. A. and Watkinson, A., 2008. Nature conservation for future sustainable shorelines: lessons from seeking to involve the public. *Land Use Policy*, 26. 203–213.

Olsen, S. B., 2003. Frameworks and indicators for assessing progress in integrated coastal management initiatives. *Ocean & Coastal Management*, 46. 347–361.

Orford, J. D., Pethick, J. and McFadden, L., 2006. Reducing the vulnerability of natural coastal systems – A UK perspective. *Managing Coastal Vulnerability: An Integrated Approach*, L. McFadden, R. J. Nicholls and E. Penning-Rowsell (eds). Elsevier Science, Oxford. 282.

O'Riordan, T., Nicholson-Cole, S. A. and Milligan, J., 2008. Designing sustainable coastal futures. *Twenty First Century Society*, 3. 145–157.

Ostrom, E., 1990. *Governing the Commons: The Evolution of Institutions for Collective Action*. Cambridge University Press, Cambridge.

Ostrom, E., 2007. A diagnostic approach for going beyond panaceas. www.pnas.org_cgi_doi_10.1073_pnas.0702288104.

Pelling, M., High, C., Dearing, J. and Smith, D., 2008. Shadow spaces for social learning: a relational understanding of adaptive capacity to climate change within organisations. *Environment and Planning A*, 40. 867–884.

Pethick, J. and Crooks, S., 2000. Development of a coastal vulnerability index: a geomorphological perspective. *Environmental Conservation*, 27(4). 359–367.

Smith, T. F., Carter, R. W., Thomsen, D. C., Mayes, G., Nursey-Bray, M., Whisson, G., Jones, R., Dovers, S. and O'Toole, K., 2009. Enhancing science impact in the coastal zone through adaptive learning. *Journal of Coastal Research*, SI 56(2). 1306–1310.

Tàbara, J. D., Cazorla, X., Maestu, J., Massarutto, A., Meerganz, G., Pahl-Wostl, C., Patel, M. and Saurí, D., 2005. *Sustainability Learning for River Basin Management and Planning in Europe, HarmoniCOP Integration Report*. Prepared under contract from the European Commission, Thematic Programme: Energy, Environment and Sustainable Development of the 5th Framework Programme 1998–2002.

Thom, B. and Harvey, N., 2000. Triggers for reform of Australian coastal management. *Australian Geographical Studies*, 38(3). 275–290.

Van der Leeuw, S., 2001. 'Vulnerability' and the integrated study of socio-natural phenomena. *IHDP Update*, 2(1), April, 6–7.

Walker, B., Holling, C. S., Carpenter, S. R. and Kinzig, A., 2004. Resilience, adaptability and transformability in social-ecological systems. *Ecology and Society*, 9(2). 5.

research paper

ENVIRONMENTAL HAZARDS
Human and Policy Dimensions

Institutional challenges to robustness of delta and floodplain agricultural systems

AUDUN SANDBERG*

Faculty of Social Science, Bodø University College, 8049 Bodø, Norway

The farming of natural coastal deltas and floodplains was in many parts of the world the cradle of civilization. Through history the transformation of risky floodplain systems into socially controlled environments was the result of an intricate interplay between ecology, demography, religion, social organization and technology of the time. After the harnessing of most large tropical rivers, there are in our time few natural deltas and floodplains left to study in the warm regions of the world. One such floodplain is the Rufiji coastal delta and floodplain in Tanzania. Here an artificial irrigation culture has not been developed, but a robust risk-minimizing system based on rice, maize, cotton and peas has evolved for delta and floodplain agriculture. Through the Arab, German and British colonization, attempts were made to 'modernize' this agricultural system, resulting in new crops and varieties being incorporated into the system in a way that made it even more robust. This study explains the genesis of this as a socio-ecological system, that is, an interaction of resource systems, resource units, governance and users. It analyses some fundamental challenges to this agricultural system during the last 100 years: the removal of the floodplain population to 'safe ujamaa villages', the recurring large infrastructure development initiatives and the modern institutional challenges such as individualized tenure, urban food market expansion, coastal and marine conservation, and the recent development of 'land-grabbing' practices.

Keywords: flood risks; institutional challenges; robustness

1. Introduction

This paper takes the reader on a journey both in time and in space. The overall idea is to provide a perspective of how coastal hazards and flood hazards – or risks – are handled in spaces different from the familiar western environment where natural risk to a large extent is eliminated and replaced by human-made risks of considerable magnitude (Beck, 1992). But to understand the dynamics of deeper changes in the nature of risks, the reader is also taken on a journey in time through the recent history of the risk management system of an agricultural people, the Warufiji, who are farming the precarious Rufiji coastal delta and floodplain in Eastern Africa. This paper is based on extensive fieldwork carried out one generation ago – in the years 1972–1974 – and a shorter revisit to the same field 30 years later to analyse some of the long-term developments. A lot of the social richness from the 1974 study is left out here, in order to highlight the long-term dynamics of the institutional challenges to this socially designed risk management system. Those interested in the full account of the still intact social organization of the Warufiji as it was in 1974 are referred to some of the sources given here (Sandberg, 1974; Havnevik, 1993).

As we shall explain below, an overarching puzzle inspiring this paper is the old formative question in sociology of the driving forces behind social evolution itself: the question of what it took to transform the risk-prone coastal delta and floodplain agricultural systems of the great civilizations into socially controlled socio-ecological systems. But such 'Grand Theory

■ *Email: Audun.Sandberg@hibo.no

doi:10.3763/ehaz.2010.0049 ISSN: 1747-7891 (print), 1878-0059 (online) www.earthscan.co.uk/journals/ehaz earthscan

Questions' will not be answered here. Instead we shall aim at testing some more modest 'middle-range theories' (Merton, 1968). One such middle-range theory is that 'robustness' – either evolved or purposely designed –will always be a trade-off between the ability to survive external disturbance and internal stress, and the system's maximum economic performance (Anderies et al., 2004). And that this compromise in a modern globalized world can undermine not only the social–ecological system, but also the productivity of the ecosystem itself. Further we shall see what we can learn from such probes in terms of the relationships between the ecological properties of the resource, the official governance system for the resource and the associated government infrastructure.

The Rufiji River coastal delta and floodplain in Tanzania is one of the few remaining large natural floodplains in the world. While most large floodplains of the tropical zones have been harnessed and turned into irrigation schemes, the floodplain and delta of this largest of the East African rivers remains naturally inundated by yearly floods and fertilized by annual deposits of silt. The yearly flushing also offsets the slow salt water intrusion from the Indian Ocean. A number of potential candidates offer themselves as explanations of the absence of a transformation of this unharnessed ecological system. Some of these are related to a series of ill-designed plans for hydropower development at the head of the floodplains from 1907 onwards. But most of these are related to the robustness of the ingenious way of farming and living in the floodplain as individual and collective solutions to the double risks of a flooded and salinated environment. As such they constitute an interesting socio-ecological system for managing ecological risks and reaping potentially high, but unpredictable benefits.

2. Coastal delta and floodplains as fertile fields for modern studies of socio-ecological systems

The study of 'Social–Ecological Systems' often requires lengthy observations, longer than most research projects can afford. Processes of nature, such as the interaction of coastal and river delta geologies, slowly form and maintain the soil structure that guarantees fertility. However, the growth and decay in ecocycles are often slow processes that may take many generations to work their way through the system. 'Robustness' of such systems is therefore a difficult phenomenon to observe and assess. Likewise, the many processes of human society (constitutional change, changes in inherited institutions and in cultural legacy, lessons from failed reforms, etc.) are usually incremental and invisible to the short-lived research project. However, policy decisions on how to alter sector policies, economic incentive structures and important institutions such as property rights are often based on short-term observations and quick assessments by nomadic consultants. They thus often threaten the long-term robustness of the interaction of the ecosystem and the social systems governing the human use of these (Ostrom, 2005). Coastal deltas and floodplains with recurring annual floods offer fertile research fields for studies of such robustness. Here social strategies, both individual and collective, show their qualities within a relatively short time span. If, for example, one disaster flood prompts the removal of peasants from their dwellings in the floodplain, this can have observable medium-term repercussions on the robustness of the agricultural system as a whole and on the ability of peasants to exercise 'ecology control' in relation to disease-bearing insects, vermin and predators of the wilderness. These sorts of balances between ecosystem processes and human modification of nature through social organization have been recurrent issues in East African history through centuries (Kjekshus, 1977; Maddox et al., 1996).

The deltas and the floodplains of the world have widely different stories to tell about the long-term robustness of social–ecological systems. These stories are as long as the history of humankind itself. The long enduring systems of the Nile Valley, the Mesopotamia and the Indus Valley were in most respects sustained 'Cradles of Civilization'; thus their basic social–ecological systems must have been fairly robust, although the tales of the ecological success and failures of these systems are multiple.

In modern times more cynical explanations of the relations between ecological and social evolution have been advocated, like the hypothesis of population pressure as a precondition for agricultural growth through the harnessing of nature (Boserup, 1965). Already Karl Marx classified the 'Asiatic mode of production' as one based on state-controlled irrigation combined with a system of royal despotism and ownership of all land. But in contradiction to the whole Marxist programme for explaining the evolution of 'western' modes of production, he offers hardly any explanation of how the transition from a 'natural' floodplain agriculture to a controlled irrigation initially took place. To other theoreticians, a strong central power or an oriental despot who combined religious and political powers has been seen as the crucial precondition for a transition from risk-prone flood agriculture to a 'hydraulic society' – a controlled irrigation farming system (Wittfogel, 1957). However, another theoretical perspective is that a despot might have been useful in organizing defence against foreign invaders, but the transition itself could perfectly well be taken care of by the peasants themselves through continuous intensification and refinement (involution) of the farming system in response to a rising population (Geertz, 1963).

It is to some extent a modern approach to regard human influences on biophysical processes as the cause of environmental problems, in spite of this being an integral part of human history throughout the last five millennia (Kjekshus, 1977; Anderies et al., 2004; Flannery, 2005). But there are also modern accounts of how human influence has enhanced biophysical processes that benefited human life and sustained strong social systems: The '2nd Nature' of the Javanese paddy terraces, of the Buhaya banana groves in Africa and of the Chinese rice–pig–fish cycles – all examples of social–ecological systems that were 'better than nature itself' (Geertz, 1963). What is the secret that made – and still makes – some of these systems robust and able to cope with external disturbances or internal stresses, while others collapsed or withered away? To an increasing extent, such long-lived resource utilization systems have been analysed with the aid of modern theoretical tools from various brands of institutional analysis (Campell, 2004; Ostrom, 2005). Armed with such tools, we shall take a closer look at a particular group of long-lived social–ecological systems, the high-risk and high-potential unregulated coastal deltas and floodplains of the world that have not yet been transformed into controlled irrigation systems.

The temptations to make sweeping generalizations about the evolutionary forces built into social-ecological systems will always be present. Therefore, it is important to warn against overgeneralizations beyond time and space or beyond particular technologies and particular ecologies. The risk-prone flood agriculture of the Nile Valley produced, in the millennia before the erection of the Aswan dam, a considerable surplus that was transformed into high cultures, religious belief systems, generous art and advanced science. On the other hand, the humanly controlled irrigation system of the Indus Valley seems to have decayed and produced a salinated and eroded environment. Each of the humanly designed farming systems, from the Nile Valley to the terraces of Java, has its distinct genesis. This means that the system (the planting cycles and planting strategies, the choice of crops and tilling methods) has evolved as a result of a unique and intricate mix of geopolitical circumstances, power relations, belief systems, available technologies and market access. When certain ways of using natural resources had proved to be useful, rules were quickly developed and this often becomes the 'common way'. It saves time, minimizes social transaction costs and reduces risk (or at least the social blame for failure) to adhere to this 'common way'. Thus organizations are often tailored to suit these 'rules in use', and institutions such as incentive structures and property rights systems evolve and become codified as frozen images of a certain 'way to do things'. The 'path dependency' of institutional development explains to a large extent the inertia found in agricultural systems. At the same time it explains the preference for some predictability in a basically unpredictable ecology facing any social system based on natural resource utilization (North, 1990). And the same path dependency also explains why sweeping generalizations are risky; every social–ecological system, every European, Asiatic or African mode of production have its own genealogy and must be analysed on its own merits. Despite the array of fashionable donor strategies for development, there are no panaceas for successful development of socio-ecological systems (De Soto, 2000; Ostrom, 2007).

Floodplains are cases of two-way risky environments. It means that this hazard works both ways. In ecologies without human control over stored water, the absence of floods usually spells drought and disaster for large populations. On the contrary, a peak flood that sweeps away everything that is planted in both low places and high places also leads to crop failure and famine. To ensure that nature delivers a 'good' medium flood with nourishing silt every year has so far been beyond the control of any social system. The initial social solution is increasingly advanced micro-adaptation to the highly variable effects of the flood, its variations across both time (from one year to the next) and place (between elevations). The alternative to this involution is investment in massive flood control works, that is, the total removal of floods, the storage of water and silt, and the overall replacement of inundated floodplain agriculture with irrigated agriculture. Any in-between design seems to be less workable.

Coastal delta areas are also fertile fields for social–ecological system research as the floods here are gradually offset by the tidal action as we get closer to the coast. Along this gradient, the hazard of large floods is diminishing, while the hazard of too small floods is increasing – due to the intrusion of salt water when there is no fresh water flushing. These different risk scenarios lead to different adaptations and to different attitudes as to what an optimal collective solution would contain. But for both coastal delta areas and large floodplains, massive flood control works seem unlikely to grow from peasant initiatives; they require the organizing capacity and the resources of an 'oriental despot', a strong state or a wealthy donor. For smaller floodplains, on the other hand, peasant cooperation in harnessing the water and constructing a socially controlled ecological system is feasible, as in the case of South Asian 'village tanks'. However, this will always depend on the existing property rights distribution, on the social capital in the communities involved and on the commitment to the local organization.

Most floodplains of the temperate or tropical world – many of these outside the ancient civilizations – have during the last 100 years been transformed into socially controlled ecological systems (mostly canal irrigation systems) by external agents. These are mostly high-profile national leaders combined with international development organizations such as the World Bank, International Development Agency (IDA) or UNDP. This makes it difficult to answer the underlying questions of what it takes in terms of social mobilization to make the transformation from a naturally flooded environment into a controlled environment and whether such collective action comes about through internal social processes or through the spread of new thoughts from other systems, either through copying or through coercion (Campell, 2004). It also makes it difficult to answer the larger question, whether such transformations are basically evolutionary or whether they are revolutionary, where the collapse of one social–ecological system carries the seeds of a new and better system. Here it is important to bear in mind the fundamental criticism of all social change theory: that these great transformations have happened in different millennia, with large differences in the level of hydraulic and farming technology, with large differences in crop availability and crop genetics and that a common cause therefore is impossible to identify. Although this is often disputed, there have also been some advances in culture, organizational knowledge and organizational capacity during the last 5,000 years: what it took to in terms of carrying through a transformation in 2500 BC South India might be very different from what it takes to initiate such a transformation in 2010 AD. To complicate the puzzle even more, we should also be aware that these historical transformations to a large extent have shaped contemporary societies, so that the science, the cultures and the institutions of the early hydraulic societies are very much part of the heritage of all modern civilizations. The underlying question therefore becomes more complex than a simple question of efficient social and ecological interrelations. The question of what were the driving forces behind the great transformations of precarious coastal delta and floodplain agricultural systems into managed irrigation systems becomes part of the grand puzzle that has occupied economic historians and sociologists for centuries. Several of the great theorists of classical social science have been intrigued by such revolutions in the basic mode of production and have consequently tried to design grand theories that could explain all past – and future – transformations. Not only Marx, but also Wittfogel, Weber and Geertz tried to design such comprehensive theories, without apparent success (Weber, 1978). In recent years, Douglass North has made new and more promising attempts to explain the driving forces in such dramatic shifts – to strive for reduced transaction costs and more efficient institutions (North, 1990, 2005).

But a second question can be asked, which can help us in diagnosing, and comparing, the working of different socio-ecological systems (Ostrom, 2007):

> Are humanly controlled socio-ecological Systems more robust than adaptive flood-plain agricultural systems?

The second question is purposely simpler and indicates that grand theories might usually be futile to pursue: the driving forces in Mesopotamia, the Indus Valley and Egypt must be understood in their own time and their own ecology and cannot be generalized across time and place. What in the third millennium drives social transformations might be very different from those drivers at the time of the Mesopotamian rulers. Different transport technologies, different market situations for agricultural products, different doctrines among development agencies (e.g. the World Bank) and new ecological knowledge might produce quite different results from the ones witnessed 3,000 years ago or 50 years ago.

All ecological systems are dynamic and in continuous change. A coastal delta and floodplain's alluvial environment with storm beaches and shifting river courses is certainly among the most dynamic ecosystems on earth. A social system here must therefore be able to show some of the same dynamics for an entire social–ecological system to survive. This means that the capacity for cooperation and the potential for collective action must be maintained within the social system and acknowledged by superior levels if the social system is nested into larger social systems (Ostrom, 1990, 2007). One of the strengths of institutional analysis is that the mechanisms for collective action in a particular setting can be analysed with a certain degree of accuracy. As we shall see below, in the fertile soil of the floodplain lies therefore one building block for a better understanding of how the interaction between the four major elements of any socio-ecological system, the resource system, the resource units, the governance system and the users, produces a certain outcome.

3. Rufiji coastal delta and floodplain agriculture as a risk management system

In the Rufiji delta and floodplain, a considerable rural population is still dependent on the rich soil and the abundant water at the same time as they face the fatal hazard of an unharnessed river. They have through the centuries developed an intricate farming system that attempts to reap the benefits of the hazardous floods at the same time as it guards against the obvious risks. This system has been amply described through the past four centuries and has, according to these analysts, shown a versatile capacity to incorporate new crops and new farming technologies (Bernadino [1606], 1962; Marsland, 1938; Sandberg, 1974; Havnevik, 1993). Approximately 150,000 people, mostly peasants, inhabit and farm the flooded plains and adjacent hills. This ecological system can be divided into nine agro-economic zones, each with its own peculiarities: the West Valley, the Flood Plain North, the Flood Plain South, the Inner Delta North, the Inner Delta South, the Coastal Delta North and the Coastal Delta South. The delta areas border on the Indian Ocean in a coastal mangrove forest of an estimated 55,000 hectares, which has ecosystem functions crucial to the entire Rufiji delta socio-ecological system. These include protection of the inner delta and floodplain agricultural areas from storm surges and wave pounding, erosion of soil by tidal currents and holding a rich biodiversity that is crucial for both farming and fishing in the entire Rufiji area.

In addition to these alluvial areas, we find the upland ecologies of the North Hill and the South Hill, which have both dryland farming areas and valuable timber areas (see Figure 1).

In addition to the 'agro-economic zones' at a particular time in the continuous formation of a floodplain, this is also a dynamic social–ecological system with both short-term and long-term dynamics. The most dynamic elements in a floodplain system are floods. Most floods will be most severe in the narrow entrance to the floodplain: the West Valley. Then a normal flood will spread out and slowly inundate the floodplain proper and the inner delta. When a flood meets the tidal effects in the coastal delta, its effects are largely absorbed by this. Even a high-volume flood will in most years be offset by the combined system of distributaries and the tidal effect of the delta. But when the soils are not soaked, an early high peak flash flood can cause considerable damage also here. On the other hand, a flood has to be of a considerable height and volume in order to inundate all the different alluvial soils of the floodplain and flush out and keep at bay the salinity in the delta and lower floodplain resulting from the tidal action. Here there are very visible short-term system dynamics at work: only with yearly flushing of fresh water can cultivation of the delta continue; otherwise the salt-tolerant coastal mangrove forest will expand inwards towards the inner delta.

The second element is more a constant factor with considerable influence on the social practices in the system. This is the rainfall that also varies considerably, but more predictably – from 1,000 mm/year in the delta to 600 mm/year in the West Valley – thus offering the best opportunities for rain-fed farming in areas where the floods make the least damage. This can also be easily observed in the form of considerably higher population densities and greater wealth in the inner delta areas than in the West Valley (Sandberg, 1974; Havnevik, 1993).

The third element in the risk management system of the Rufiji is the varied and slowly changing soil conditions and micro-differences in elevations in the floodplain. These are largely a result of thousands of years of flood and tide action and of differential silting as the meandering changes the course of the river. These are long-term dynamic alluvial processes that lead to a very complex micro-ecology with layers of soil types sandwiched in different ways all through the floodplain. River courses change continuously while old river levees and ox-bow lakes are left behind. Even old storm beaches from coastal action far into the delta are part of this complexity. The lowest-lying parts of the floodplain have largely clay soils, *Kitope*, which have high soil moisture retention capacity and poor drainage and are hard to work with hand tools. The most preferred soils, the *Mbaragilwa* soils, are found at medium elevations in the floodplain, on old river levees and other areas that are normally lightly flooded. These are mixed clay and sandy soils. They usually have good water retention capacity and good drainage and are easy to work with hand tools. The *Gongo* soils are found on the highest river levees not flooded during a 'normal' flood. These are mostly sandy soils, they have poor water retention capacity and

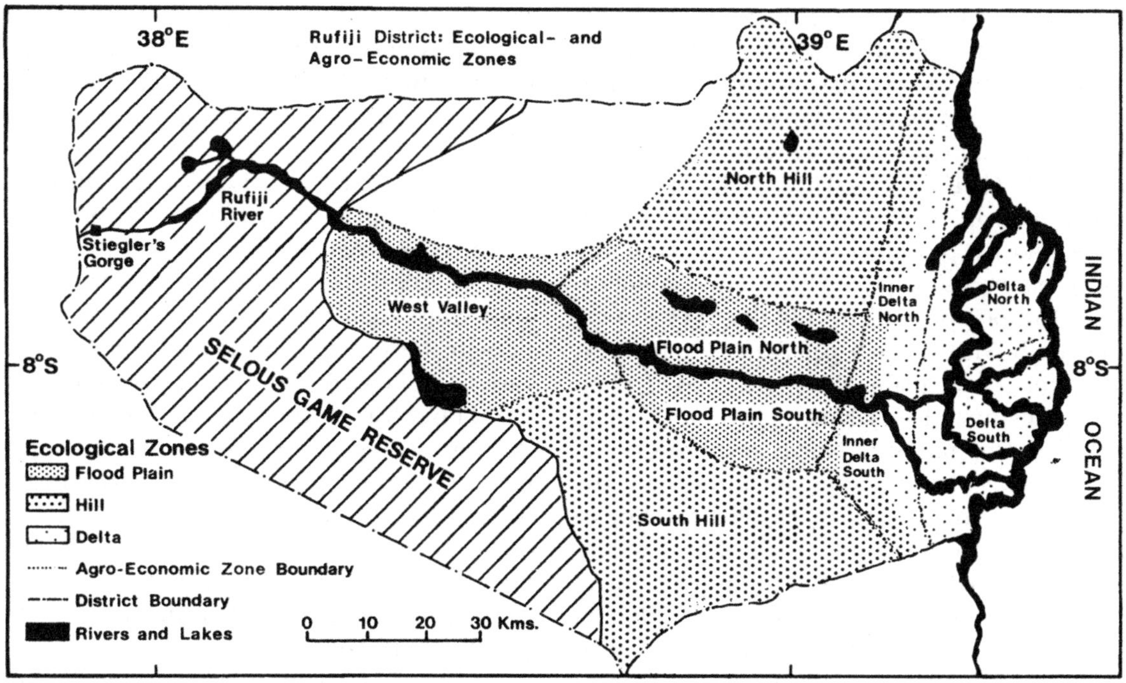

FIGURE 1 Agro-economic zones of the Rufiji District, Tanzania
Source: Sandberg, 1974.

good drainage, and they are easy to work with hand tools. But in addition to these three major ecosystem properties, experienced farmers also view the ecosystem in a three-dimensional and dynamic perspective. They distinguish between uniform alluvial clay and clay overlying *Mbaragilwa* or sand, as well as between uniform *Mbaragilwa* and *Mbaragilwa* overlying sand or clay. In a dynamic perspective, the users of the preferred *Mbaragilwa* soils 'need' a substantial flood with considerable silt deposit every third year. If this does not happen, the yield will drop to about one half without the application of artificial fertilizers (Sandberg, 1974; Havnevik, 1993).

Through the centuries there has evolved among the peasants of Rufiji an intricate agricultural system that seems to minimize risks and maximize benefits as far as this is possible. The system has evolved from the bottom up, not only as peasants' response to the ecological factors mentioned above, but also as a social response to various colonial and state interventions from above into rural production. As we shall see below, this also includes social responses to new technological and marketing possibilities. It thus represents a social–ecological system characterized by continuous change. There is a long discussion on whether this system at any time has been at 'its peak equilibrium' and subsequently decayed due to external disturbance and the influence of modernity (Marsland, 1938; Sandberg, 1974; Kjekshus, 1977; Havnevik, 1993). It is important to understand, however, that to most peasants, the 'system' as such means little; it is the farming strategies open to them that produce their individual and collective choices and thus reproduce at every new season what at a given time can be observed as a floodplain agricultural system. Therefore, the evolutionary history of the agricultural system as such is not so relevant to peasants in this perspective, apart from the path dependency that particular ways of doing things in the past create for future generations. In analysing the robustness to flood and coastal hazards, however, a system perspective is highly relevant. The research question above can thus be rephrased and sharpened in the following way: with nine agro-economic zones, 8–10 different crops, 3 × 3 soil variations and complex multitudes of micro-elevations, should there not in principle be sufficient strategies available for any peasant to meet any perceivable flood scenario? And if so, should not this floodplain agricultural system constitute a 'robust' social–ecological system with no need for transformation into an irrigated ecological system? Or are there other stress factors that challenge the robustness of the system? And are these always external, or are there also inherent social factors that weaken this robustness? To answer these questions, it is necessary to look more in detail at the agricultural system with crops and farm labour input, as it was practised in the second half of the 20th century.

Conveniently the Rufiji agricultural system could during the 1970s be seen as consisting of two separate subsystems: the century-old *Mlau* system – the dry season farming system (June–September), and the newer *Mvuli* system – the flood season farming system (February–June). The Mlau system is a low-risk system, a flood recession planting strategy where the roots of the growing plants follow the sinking moisture level after the flood is over and all parameters (flood heights, duration and soil properties) are then known at the time of planting. The Mvuli system, on the other hand, is a high-risk system facilitated by the opening of a new market for rice by the Zanzibar Sultanate about 400 years ago. Here the properties of the coming flood were unknown and the farmers had to prepare for alternative flood scenarios by planting different rice varieties and at different elevations. The typical farmers' calendar would thus look like Figure 2 for a farmer who tries to implement the entire two-pronged system.

In detail, the farmer's calendar consists of a number of choices at regular intervals through

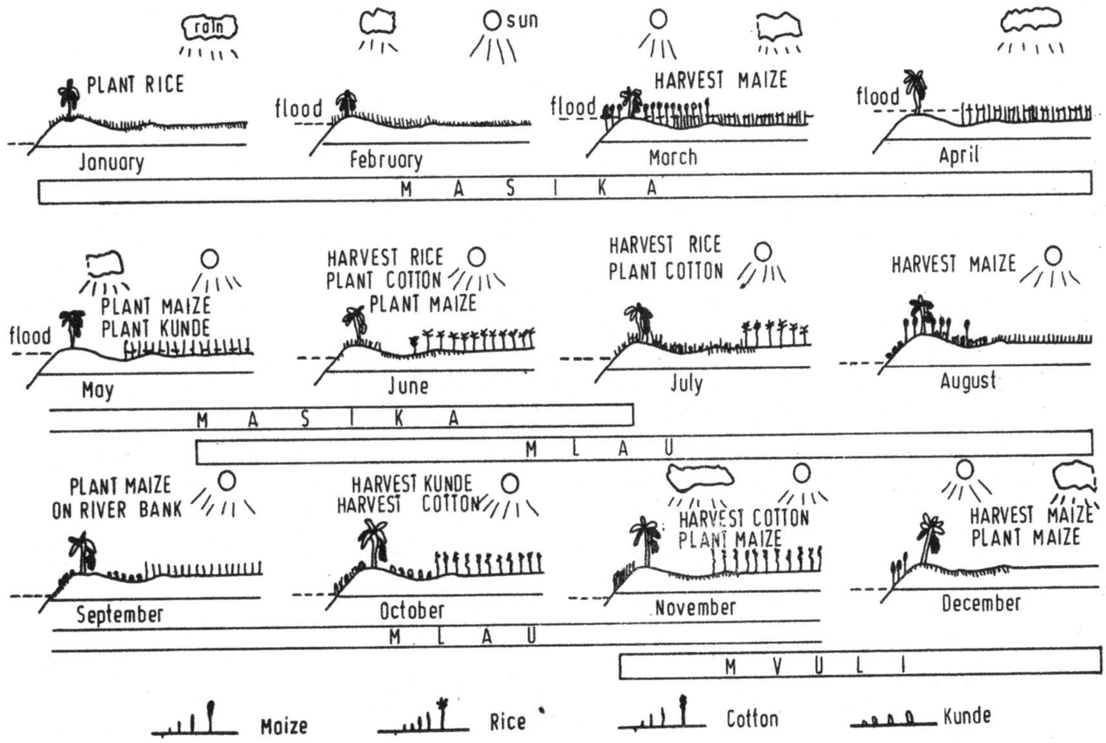

FIGURE 2 Farmer's calendar 1974 – Rufiji Flood Plain
Source: Sandberg, 1974.
Note: *Kunde* = cowpea leaves.

the year. The choice of every month will to a large extent depend on the observed prospects for the harvest of the crops planted a few months earlier. Figure 3 sums up all the alternatives open to a typical floodplain farmer and the main considerations that enter the farmers' choice considerations.

Within the given ecological system, this constitutes a sophisticated system of allocation of human resources throughout the year. Although it is virtually impossible for every peasant to employ all the strategies for the entire system within one year, it is important to think of the system as a set of collective strategies at the disposal of all members; in short, it is part of the Rufiji peasant culture. Although this might not result in an optimal system in terms of output per man-hour spent in farming, we can conclude that the system is robust in the face of the risks involved in flood farming. This robustness also comes from the flexibility of the system and its transparent feedback mechanisms. If the system is disturbed by a shock-like absence of rain or a devastating flood, it is self-correcting within the same agricultural year; thus it also has a high resilience. If a high and voluminous flood drowns all the rice in April, there are no labour peaks in June, but large areas of moist and fertile soils that can be planted with cotton, maize and cowpeas. Such large floods tend to happen with 4–7-year intervals – during the last century large floods were recorded in 1930, 1936, 1940, 1942, 1944, 1945, 1952, 1956, 1962, 1968, 1974, 1979 and so on (Havnevik, 1993). If the flood is medium or small, the rice yield is usually good provided that the right rice variety had been planted in the right micro-environments (Sandberg, 1974). In such cases there are neither vast areas nor many free hands available to plant large quantities of cotton and maize.

As explained above, the Rufiji coastal delta and floodplain agricultural system of the 20th century could be seen as a fully evolved risk-minimizing system in which farmers absorb all the risk themselves. As such it is a robust social–ecological system capable of accommodating most environmental uncertainties. The farmers of Rufiji have also shown a capacity to include the right new

Month	Flood level	Available strategy	Output/constraint
November	Low water	Plant maize on high river banks	Good crop if low flood
		Start preparing rice fields	Avoid labour bottlenecks
		Harvest and market cotton	Mlau system
		Harvest and market mango fruits	
December	Low water	Harvest maize from low clays	Mlau system
		Preparing rice fields continue	Avoid labour bottlenecks
		Plant maize on high river banks	Good crop if low flood
		Harvest and market mango fruits	
January	Low water	Plant rice, fast/slow > high/low	Right variety in right place
February	Rising water	Rice is growing	
March	Rising water	Rice is growing	Good crop if medium flood
		Harvest maize from river banks	Needs dry spells to ripen
April	Flood peaks	Rice grows	Tolerates medium floods
May	Flood recedes	Rice ripens	
		Plant maize on highest places	As flood recedes
		Plant cowpeas on highest places	As flood recedes
June	Flood recedes	Plant maize on *mbaragilwa*	As flood recedes
		Plant cotton on *kitope*	As flood recedes
		Harvest rice	Labour peaks
July	Flood recedes	Harvest and market rice	
		Plant cotton on lowest clays	
August	Low water	Harvest maize from high places	
September	Low water	Plant maize on lowest clays	
October	Low water	Harvest cowpeas	
		Start harvesting cotton	

FIGURE 3 Details of farmer's calendar

crops and new technologies into their strategic repertoire: rice during the Zanzibar Sultanate, maize from South America, cotton during the German Colonial Period, cashew nuts during the Ujamaa period, together with a gradual expansion of oranges and mango orchards. All these crops were incorporated into the agricultural system in a way that strengthened the system and contributed to make it gradually more robust as a self-contained risk management system. Most of this incorporation took place as individual adoption of new planting and labour strategies, often after collective resistance to the forced introduction of the same crop as a new 'cash crop'.

But as it takes a lot of human concern and effort to prepare for several possible, but unknown future flood situations, the economic return to labour will always be low in such a system, despite its apparent robustness. It takes a lot of extra work to constantly prepare for multiple futures. Low labour productivity is therefore the premium the Rufiji peasants have to pay for their complicated risk insurance system. As mentioned above, a maximum economic efficiency is also not to be expected from a complex and robust social–ecological system. A precondition for the ecological robustness has therefore been the low labour cost characterizing Rufiji as a typical rural area in Tanzania. Faced with modern market forces resulting from different donor-driven structural economic reforms, in crop markets, in capital markets, and eventually also in labour markets and in land property markets, this low labour productivity represents a new kind of challenge to the system's robustness. This became increasingly more pronounced towards the end of the 20th century and at the beginning of the 21st century. Now, with new development scenarios where such risks are likely to be absorbed by others than the peasants themselves, for example, by financial markets, new farming strategies with higher labour productivity are likely to take precedence over traditional risk-management strategies. This will in a profound way affect the nature of the robustness of the Rufiji as a self-contained social–ecological system.

The reason for this is that the sustained robustness of the traditional agricultural system itself depends on a number of conditions that have to be met. They are mostly related to the micro-environmental knowledge among the peasant population and especially the dynamic aspects of this:

- shared knowledge in the floodplain communities of local diversity of flood heights and onset times in various fields
- shared knowledge of drainage and fertility properties of the various soils
- shared knowledge of place and timing of vermin attacks
- shared knowledge of the flood resistance properties of seed varieties, especially of rice
- shared knowledge about the importance of traditional land tenure mechanisms with continuous redistribution, which allow most farmers to make the right strategic choice in relation to time and place for planting and use of human energy.

If these preconditions are not fulfilled, that is, actively maintained and supported, this will affect the long-term robustness – or sustainability – of the traditional system. Among the conditions from which such robustness is likely to be negatively affected is for example, the breakdown of effective transmission of knowledge between generations and land tenure changes initiated without the participation of the peasants. Recent developments in Tanzania show a number of examples of cases that have a deep impact on the robustness of intricate traditional risk management systems like the Rufiji coastal delta and floodplain agriculture. We shall briefly deal with some of these institutional challenges below.

4. Ecology control initiatives

The first state effort to modify the social–ecologial system of the Rufiji floodplain was the forced cultivation of cotton introduced by the

German Colonial Government in 1902. It threatened and interfered with the traditional rice and maize system evolved during the Zanzibar Sultanate and led to the largest peasant uprising in Colonial Africa, the *Maji Maji* rebellion (1905–1907) where altogether 75,000 Africans were killed. However, after the rebellion was over, the Rufiji peasants individually adopted cotton as a cash crop and planted large areas with cotton under their own control, especially when the Germans established ginneries and efficient market outlets for the crop (Havnevik, 1993).

But new crops were not always the result of unilateral state initiatives. Apart from a massive cashew nut campaign during the *ujamaa* period, most new crops, like mango and orange, were adopted gradually without any campaigns and incorporated into the agricultural system explained above so as to transform this into an even more intricate social–ecological system.

After a devastating flood in 1968, the Tanzanian Government decided to alter the social–ecological system of the Rufiji dramatically. Various objectives like future disaster prevention and the ideological aim of 'living and working together' were mixed to produce the first *ujamaa* campaign in Tanzania in 1968–1972 (Nyerere, 1968). The resettlement process was carried out without real participation by the inhabitants of the floodplain and considerable force or 'heavy persuasion' was applied. This collectivization campaign removed the majority of the peasants from their *madungu* dwellings (houses on poles) in the floodplain and relocated them in 'secure planned villages' on higher ground. By 1973, 75 per cent of the population in Rufiji were living in 25 *ujamaa* villages; the remaining 25 per cent were either not affected by the operation (the Coastal Delta North and the Coastal Delta South) or refused to move because the proposed villages were too far from their fields (Inner Delta South). In the West Valley the peasants lost most of the ecological control over their fields, as the new villages were established on the north side of the Rufiji River while their fields in the floodplain valley remained on the south side of the river, now bordering directly on the Seleous Game Reserve. Without any human settlement as a barrier against wild animals, the attacks on crops by elephants and wild pigs added new risks to the difficulties of farming in this part of the floodplain, and a gradual depopulation started (Sandberg, 1974; Ndunguru and Hahn, 1998). Even if the national policies of resettlement and collectivization were given up in the 1980s and 1990s, the resettlements were not reversed and the social processes set in motion in the early 1970s have continued into the 21st century.

In the Flood Plain North the distance to the floodplain fields was short; here the peasants in the new villages could continue the agricultural system explained above more or less uninterrupted after 1973. But here, other social mechanisms started to work as a result of the massive resettlement. On the one hand, a gradual urbanization of these villages has increased the demand for and the value of good adjacent floodplain land. On the other hand, the growth of these *ujamaa* villages, especially Ikwiriri, now a town of 40,000 inhabitants, has facilitated specialized occupation and a growth in local food markets. This has altered the strategic options for peasants: staples and cereals for sale to the State Marketing Board are no longer the only options. A wide range of different grains, fruits and vegetables are now demanded by an increasing number of more sophisticated urban consumers within the Rufiji district itself. The most industrious peasants have thus gradually turned into farmers.

In 2003 a new major bridge across the floodplain and river from Ikwiriri was completed. This has dramatically increased the marketing possibilities for all crops from the whole of the Rufiji floodplain. Following the resettlement programme, the reluctant villagers in the Flood Plain South and in some other parts of Rufiji were reported to maintain the traditional floodplain system somewhat throughout the 1980s and 1990s by establishing peripheral houses in their old fields, thus commuting between the new villages and the floodplain and keeping to the old ways (Havnevik, 1993). With the new bridge, these are now also within easy reach of the markets in Ikwiriri and Dar es Salaam. So with the exception of the villagers of the West Valley, who have had to receive famine relief, the total effect of the resettlement of farmers from the floodplain to higher ground villages has therefore over the decades not destroyed the social–ecological system evolved in the Rufiji. But it has led to some major changes in the choice of crops and planting strategies, especially in the Flood Plain North, and has in a dramatic way brought the farmers more into the mainstream of national development.

Many attempts have also been made to change the very fundamentals of the ecological system of Rufiji itself by harnessing the Rufiji River and introducing flood control as engineered solutions. This would be done in the shape of plans for a large dam at Stiegler's gorge for the combined purpose of hydro-electric power production and flood control. Already in 1907 the Germans (Stiegler) surveyed the Rufiji basin for the potential for irrigated agriculture and hydropower generation. Surveys in 1928, 1952 and 1961 (FAO) were based on a 'multipurpose' idea of replacing the flood agriculture of Rufiji with an artificial irrigated agriculture combined with absolute flood control and hydropower production. However, the FAO report concluded that the cost to each acre of net irrigable land (flood control and irrigation) would be rather high, and that more than half of the costs therefore should be charged to hydropower generation. Subsequent surveys in 1967 (USAID), 1968 (JETRO), 1972 (NORAD/Norconsult) and 1980 (NORAD/Hafslund) therefore put most emphasis on the hydropower generation potential. Gradually the criticism against this single-purpose planning strategy grew and a multipurpose approach was reintroduced. But contrary to the early German ideas of a massive transition from flood cultivation to artificial channel irrigation, this approach was a modified flood release design, which aimed at not losing too much electricity, while releasing sufficient floods through low-level outlets to allow the floodplain agriculture to continue during a lengthy transition period towards fully irrigated agriculture of an undisclosed nature. This transition was however not part of the project and there was only minimal participation of the Rufiji peasants in the design of the planned mega-project. The multipurpose compromise was therefore a poor one, so that the feasibility of both the hydropower objectives and the agricultural objectives was only marginal. In addition, the measures to combat the slow salt intrusion on agricultural land were not convincing. In the end, the World Bank turned down the project; only a comprehensive multipurpose project would stand a chance at the World Bank. So all these planning efforts during a period of 100 years resulted in nothing, and in many ways left both the floodplain ecosystem and its robustness intact. There might be many explanations offered for this distinct non-transition to irrigated agriculture during the whole of the 20th century. Among such explanations are the particularities of this floodplain ecology, the internal politics of Tanzania, the role of donors, the expertise of the chosen consultants and the understanding of sector bureaucrats. The peasants of Rufiji have, however, largely been absent as stakeholders in the 100-year discussion and planning of hydropower development and of a possible total change to their social–ecological system. If the issue of flood control, power generation and irrigated agriculture should appear again in the 21st century, farmers might now become a more integrated part of the planning and implementation process. But as we shall see below, these farmers are most likely to be quite different from the typical risk-managing peasant of the 19th and the 20th century.

The mangrove forests in the Coastal Delta have been utilized as a source of building materials for many centuries. Elevated houses on poles, in Rufiji, Malindi, Lamu, Hadramuth and Oman, were largely built on Rufiji mangrove poles. This logging activity was largely sustainable due to a low level of technology and later due to state regulations. But in the mid-1990s, plans were launched by the African Fishing Company (AFC) to cut down 1,200 hectares of the Rufiji mangrove forest in order to build prawn ponds and create the world's largest shrimp aquaculture facility with a projected US$500 million profit per year. Despite strong opposition from local Rufiji people, the Tanzania National Environmental Management Council (NEMC) and numerous international environmental non-governmental organizations (notably IUCN and WWF), the Tanzanian Cabinet approved the project. However, due to continued local and international resistance, the implementation of the project was delayed (Bryceson, 2002). Huge debt accumulated and in 2001 the AFC was liquidated, thus putting an end to the shrimp project. Thus the Rufiji coastal mangroves were saved, and the IUCN could bring to fruition the joint environmental management plan with the Rufiji District Council (REMP) that they had been working on since 1997. Already in 2005 the Rufiji coastal delta and its mangrove forest were thus officially protected as part of the Rufiji–Mafia–Kilwa under the protection of the International Ramsar Convention. The objective of this is 'the long-term conservation of the forests, woodlands and wetlands, so that biodiversity is conserved, critical ecological functions are maintained, renewable natural resources are used sustainably, and the livelihoods of the area's inhabitants are protected and enhanced' (IUCN, 2005). Partly in parallel, the WWF Africa and Madagascar Programme has initiated a Rufiji–Mafia–Kilwa Seascape programme that represents an 'eco-regional' approach to sustainable natural resources management and biodiversity conservation. The stated purpose here is the 'Improved socio-economic well-being of coastal communities in Rufiji, Mafia and Kilwa districts through sustainable, participatory and equitable use and protection of their marine and coastal natural resources' (WWF, 2009).

Thus, in the course of only 100 years, the Rufiji coastal delta and floodplain inhabitants have been exposed to widely different ideas of modernization, from the traditional heavy engineered solutions for the harnessing of rivers for total flood control – and massive shrimp aquaculture – to the most modern resilience ideas of natural floodplain forest and mangrove ecosystems as suppliers of cost-effective services to coastal and riverine human societies: storm protection, coastal defences, carbon sequestration, water recharge and nurseries for local fisheries (UNEP, 2010; WB, 2010). But neither of these development strategies has so far changed much of the social fabric of the Warufiji communities; they still practise a 'living with the floods strategy'. Then what, if anything, is likely to affect the robustness of the coastal floodplain agricultural system?

5. Land tenure changes and modernization of the social system

The struggle of the Warufiji, from the early colonial times and up to 2009, has taught them the lesson that their land in the floodplain is valuable and worth fighting for. The potential for collective action was traditionally tied to the traditional settlement pattern. Here the fields were organized around the homestead, with individual family members, including wives and in-laws, being allocated their own plot for the current growing season by the family head. These patterns we could still find in the 1970s, especially in the Inner Delta South. A typical homestead here consisted of 20–30 people farming about 4–6 hectares continuously – in various locations and elevations and on different soils. The micro-ecology of the area decided to a large extent which crops could be grown when and in what place, and the working ability of the individual decided to a large extent the size of a particular plot (Sandberg, 1974).

During the *ujamaa* programme, the state launched a campaign for massive collective farming on larger fields. However, the peasants had already had bad experiences with the cooperative farms of the corrupt and defunct Rufiji Co-operative Society and did not see collective farming as an attractive novelty. The solution devised by the new village collectives was that communal farms could be demarcated on the infertile and sandy higher grounds close to the new villages, while the land tenure of the fertile floodplain was left intact. As no annual crops would succeed on the higher ground, they were planted with cassava and a new crop, cashew nut trees, which, if tended properly, can give a considerable cash income after a number of years. Thus the effects of the period of collectivization in Tanzanian agriculture were hardly noticeable in floodplain agriculture.

However, after four decades the effects of the physical relocation of people to *ujamaa* villages have gradually become more noticeable. As the demographic processes of ageing, marriages and birth of new generations worked their way into land use, the land tenure of the floodplain did in practice gradually become more and more fragmented in relation to the new settlement pattern. One person can now have different plots through different family affiliations in widely different places. The advantage of this is that risks are still low if these plots are in different micro-environments. The disadvantage is that labour productivity in tilling, planting and harvesting is even lower than before, due to the need for extensive commuting between the village and the various plots. Also the institutions for allocation of plots to new members of villages (youngsters, married or migrants) come under pressure when fragmentation becomes too intricate and the customary village land property rights start to erode from within. Some effects of more cumbersome land tenure arrangements can be seen in the increasing number of young men engaging in cash-generating 'bizzniss', like charcoal trade, illegal hardwood logging and fruit sales (mango and orange) to urban markets in Dar es Salaam. These give higher and more immediate returns to labour than continued cumbersome farming of staples in the floodplain. For instance, the price of mango in 2003 was reported to increase 10 times by being moved 200 km from the market in Ikwiriri to the market in Dar es Salaam. But apart from the new attractions of profitable trade, there would in principle not have been any hindrance for a continuous adjustment of land tenure by a local land consolidation process through local land tribunals at the district level – with active involvement of the local collectives in Rufiji villages. But such processes did not take place, mostly because the central government from 1970 to 1990 gave no active support to such local-level land tenure processes. In a nested system of land tenure processes, the national deadlock on national land tenure policies severely hampered the necessary continuous local adjustments of land tenure to changing demographic circumstances.

Under pressure from the World Bank for accelerated 'structural adjustments', Tanzania embarked on a road 'from *ujamaa* to economic liberalization' during the 1980s (Gibbon et al., 1993). As part of this, a new agricultural policy was formulated, which opened up for 'institutionalizing a market for land' – also in rural areas. A long struggle took place during the 1990s over the question of whether village land property rights should be codified in new legislation (United Republic of Tanzania, 1992; The National Land Policy, 1995), or whether it should only be protected administratively without any title, but with freedom for villagers to enter into land transactions also (McAusland: Draft Bill for the Land Act, 1996; Izumi, 1998). Finally, and still under pressure from the World Bank, a Land Act that sought to establish a market for agricultural land was passed. This enabled villagers to buy land and get a title, in line with some of the new land policies in a number of African countries that had been inspired by Hernando de Soto's ideas of combating poverty by giving secure property rights to the poor (De Soto, 2000). But these policy changes will also enable foreign investors and political dignitaries to buy large tracts of land for more efficient mechanized commercial agricultural production than the customary peasant agriculture could demonstrate (Ikdal, 2000; Lindstrøm, 2001). So-called 'land grabbing', that is, the conversion of political capital into individual land property, is now legally possible under the new act. The exchange of political favours towards special villages against subsequent individualization of village land is only one way of making such conversions. Also, foreign capital can now be converted into property rights to fertile tropical land, as the case has been with capital from the oil-rich Gulf states invested in the Ethiopian highlands.

The Rufiji floodplain has for hundreds of years been a very attractive tropical agricultural area, and rich villagers in Rufiji with external sources of income, foreign food companies, and national industrial and national political leaders will be interested in acquiring individual property rights in this area. These are not forced to employ the intricate risk-managing strategies of the traditional Rufiji peasant, but can concentrate on the relative risk-free *Mlau* after-flood cultivation strategy for large-area mechanized farming. An increasing number of entrepreneurial villagers from the most urbanized villages now maintain two houses, one in the village and one in one of Dar es Salaam's suburban areas. Thus the village collectives of the 1970s and the older homesteads based on kinship are gradually eroded and the potential for this kind of collective action is decreased. The new bridge across the Rufiji River has also 'opened up' the vast area of floodplain south of the river for non-subsistence agricultural development and has brought the whole district much closer to the market. In addition, the market itself has expanded extremely rapidly. The national capital Dar es Salaam has grown from a population of 0.75 million in 1974 to 3.5 million in 2003. This market now works as an enormous stimulus for accelerated commercialization of agriculture in the Rufiji floodplain. Especially in fruits and vegetables there is an insatiable market in the whole metropolitan area of Dar es Salaam. Fruit trees, such as mango, thrive in the floodplain and are robust against floods and droughts; also the labour input is minimal. However, embryonic, labour-intensive pump-irrigated vegetables also have a high potential profit in the urban market provided that the logistics of rapid transport and efficient marketing is in place. The social processes in the Rufiji agricultural communities resulting from these new market openings will take several decades to have full impact, but will most probably further strengthen the importance of individualized property rights to coastal delta and floodplain land.

6. Implications for the robustness of coastal delta and floodplain social–ecological systems

Changes in land tenure do not alone affect the robustness of a social–ecological system. It is quite feasible to have a gradual change in land tenure with maintenance of robustness. But changes in agricultural practices resulting from the interaction of new and more profitable marketing openings and rapid individualization of land tenure can dramatically influence the robustness of a system. A peasant risk management system can, as we have seen above, be robust against external disturbances for a long period, incorporating new crops and expanding the array of peasant strategies. But this often happens at the price of hard toil and labour bottlenecks. It is therefore possible that this traditional robustness could have been strengthened and prolonged by introducing micro-credit schemes, especially for women cultivators, to overcome the financial bottlenecks that lower the efficiency of the system and to prevent short-term famines.

But with the development towards increased individualized tenure and increased commercial agriculture in Rufiji, the social–ecological system is likely to go through a qualitative change. This can be summed up in a 'slow movement from peasants to farmers'. When the burden of risk-taking shifts from the local peasant community to the national banking system or to multinational corporations, the social–ecological system has qualitatively changed and the criteria for socio-ecological robustness have to be changed correspondingly. Now a single after-flood contract crop (*Mlau*) with high market potential can catch a high value at a low risk. And combined with perennial flood-resistant mango fruit trees and pump-irrigated intensive vegetable cultivation in the dry season, these strategies can be elements in a completely new type of social–ecological system, qualitatively different, but still robust. This would have a higher degree of specialization and a much more unequal social fabric than the typical Tanzanian peasant communities of the 20th century. With a larger portion of the risk located outside the floodplain, it would represent another stage in the transition of agricultural systems in coastal floodplains. For analysts of socio-ecological systems, varying future institutional developments can also be analysed with the diagnostic tools developed within the social–ecological system literature, where the interactions between the resource system, the resource units, the governance system and the users produce outcomes that can be evaluated in terms of robustness and maintenance of crucial ecosystem services (Ostrom, 2009). There is no panacea for a particular way of floodplain farming as an ideal risk management system, but we have shown here that in such analysis the existence of a coastal hazard and risk management system evolved by the users themselves are fertile fields for applying these kinds of analysis.

Thus, with the potential of a growing commercial agriculture sector in the Rufiji floodplain, combined with modern conservation measures protecting the coastal delta mangrove forest and its beneficial ecosystem services, the questions at the beginning of this paper have to be rephrased.

The initial 'grand theory' question of the preconditions for a transition from flood-prone agriculture to large-scale artificial irrigation in the 21st century will be far less relevant than what it was in the 20th century. This is because the balance of power between the state and its foreign donors, on the one hand, and the state and the farmers, on the other hand, is likely to shift in favour of the more commercial farmers and the environmental non-governmental organizations. The policy implication of this is that these groups, both through stakeholder representation and through political and corporate channels, will demand a much greater say in how the social and political control over the crucial coastal delta and floodplain ecosystems shall be designed than was the case in the 20th century. The future of the Rufiji social–ecological system is therefore more open than ever, and only the rest of the 21st century can show in what form its inherent robustness will manifest itself.

Acknowledgements

This paper was initially presented for the theme group 'Historical Institutional Analysis on Robustness of Social-Ecological Systems', at the WOW 3, Workshop on the Workshop, Indiana University, 2–6 June 2004. It has since undergone considerable revision and shortening. A travel grant from the Norwegian Academic Authors' Association to facilitate a revisit to Rufiji in 2003 is highly appreciated.

References

Anderies, J. M., Janssen, M. A. and Ostrom, E., 2004. A framework to analyze the robustness of social-ecological systems from an institutional perspective. *Ecology and Society*, 9(1). 18.

Beck, U., 1992. *Risk Society: Towards a New Modernity*. Sage, London.

Bernadino, G. D. S., 1962. *A Journey along the Coast in 1606, The East African Coast, Selected Documents from the First to the Earlier Nineteenth Century*. G.S.P. Freeman-Grenville, Clarendon Press, Oxford.

Boscrup, E., 1965. *The Conditions of Agricultural Growth, the Economics of Agrarian Change under Population Pressure*. Aldine Publishing Company, Chicago.

Bryceson, I., 2002. Coastal aquaculture developments in Tanzania: sustainable and non-sustainable experiences. *Western Indian Ocean Journal of Marine Science*, 1(1). 1–10.

Campell, J. L., 2004. *Institutionel forandring og globalisering*. Akademisk Forlag, Copenhagen.

De Soto, H., 2000. *The Mystery of Capital*. Basic Books, New York.

Flannery, T., 2005. *The Weathermakers*. Text Publishing, Melbourne.

Geertz, C., [1963] 1970, *Agricultural Involution, the Processes of Ecological Change in Indonesia*. Univerity of California Press, Los Angeles.

Gibbon, P., Havnevik, K. J. and Hermele, K., 1993. *A Blighted Harvest – The World Bank and African Agriculture in the 80s*. Afrika World Press, James Curry, London.

Havnevik, K. J., 1993. *Tanzania, the Limits to Development from Above*. Nordiska Afrikainstitutet, Uppsala.

Ikdal, I., 2000. *Human Rights and Rural Women's Land Rights, An Analysis of the Land Tenure System and the Land Reform in Tanzania*. Undergraduate dissertation for the law degree, February 2000, University of Oslo (Ins. For Kvinnerett Univ i Oslo).

IUCN, 2005. www.ens-newswire.com/ens/jan2005/2005-01-28-04.html.

Izumi, K., 1998. *Economic Liberalisation and the Land Question in Tanzania*, Phd dissertation, International Development Studies, Roskilde University, Denmark.

Kjekshus, H., 1977. *Ecology Control and Economic Development in East African History, The Case of Tanganyika 1850–1950.* Heineman, James Curry, London, 1996.

Lindstrøm, E., 2001. *Human Rights and Rural Women's Land Rights – An Analysis of the Land Tenure System and the Land Reform in Tanzania.* Studies in Women Law n. 44, Institutt for Offentlig rett, nr. 4/2001, University of Oslo.

Maddox, G., Giblin, J. L. and Kimambo, I., 1996. *Custodians of the Land, Ecology & Culture in the History of Tanzania.* James Curry, London.

Marsland, H., 1938. *Mlau-cultivation in the Rufiji Valley.* Tanganyika Notes and Records V, pp. 55–59.

McAusland, 1996. *A Draft of a Bill for The Land Act.* Prepared for The Ministry of Lands, Housing and Urban Development, United Republic of Tanzania, under an ODA contract.

Merton, R., 1968. *Social Theory and Social Structure.* Free Press, New York.

Ndunguru, I. F. and Hahn, R., 1998. *Reconciling Human Interests with Conservation in the Selous Game Reserve (Tanzania).* Paper given at the International Workshop on: Community-Based Natural Resource Management (CBNRM), World Bank, Washington, 10–14 May.

North, D. C., 1990. *Institutions, Institutional Change and Economic Performance.* Cambridge University Press, New York.

North, D. C., 2005. *Understanding the Process of Economic Change.* Princeton University Press, Princeton, NJ.

Nyerere, J., 1968. *Uhuru na Ujamaa, Freedom and Socialism: A Selection from Writings and Speeches: 1965–1968.* Oxford University Press, Dar es Salaam.

Ostrom, E., 1990. *Governing the Commons – the Evolution of Institutions for Collective Action.* Cambridge University Press, New York.

Ostrom, E., 2005. *Understanding Institutional Diversity.* Princeton University Press, Princeton, NJ.

Ostrom, E., 2007. A diagnostic approach for going beyond panaceas. *Proceedings of the National Academy of Sciences*, 104(39). 15181–15187.

Ostrom, E., 2009. A general framework for analyzing sustainability of social_ecological systems. *Science*, 325. 419.

Sandberg, A., 1974. *Socio-Economic Survey of Lower Rufiji Flood Plain – Rufiji Delta Agricultural System.* BRALUP Research Paper No. 34, University of Dar es Salaam.

United Republic of Tanzania, 1992. *Report of the Presidential Commission of Inquiry into Land Matters*, Vol. 1, Land Policy and Land Tenure Structure. November.

UNEP, 2010. C. Nellemann and E. Corcoran (eds). *Dead Planet, Living Planet – Biodiversity and Ecosystem Restoration for Sustainable Development.* A Rapid Response Assessment, United Nations Environment Programme, GRID-Arendal. www.grida.no

Weber, M., 1978. *Economy and Society.* University of California Press, Berkeley, CA.

Wittfogel, K. A., 1957. *Oriental Despotism: A Comparative Study of Total Power.* Yale University Press, New Haven, CT.

World Bank, 2010. *Environment Matters at the World Bank – Banking on Biodiversity.* IBRD, Washington DC.

WWF, 2009. Rufiji-Mafia-Kilwa Seascape Programme, Tanzania, http://assets-panda.org.

research paper

ENVIRONMENTAL HAZARDS
Human and Policy Dimensions

Strategic regeneration: A policy coupling approach to managing a coastal resort in South Wales

DEBORAH PEEL* AND GREG LLOYD

Built Environment Research Institute, School of the Built Environment, University of Ulster, Shore Road, Newtownabbey, Co. Antrim BT37 0QB, Northern Ireland

This review paper takes a case study approach to critically discuss how regeneration in the coastal resort of Porthcawl, South Wales has integrated flood defences as part of a strategic spatial planning and integrated policy coupling approach to the redesign of the town's seaside, promenade and harbour quarter. This planned initiative is part of a concerted effort to upgrade the town's waterfront and enable regeneration of a small seaside resort. From a resort cycle perspective, Porthcawl's perceived vulnerability to changing economic fortunes has prompted a master planning approach which is intended to bring together a mixed-use development programme with particular attention to urban design, strategic environmental assessment and public consultation. In different ways, this process has sought to raise public awareness of the issues and risks associated with the town's coastal location and to integrate flood defence aspects within the strategic regeneration plan as part of a greater sensitivity to climatic risks and potential flooding. This paper examines the adaptive capacity of coastal communities and the complex of socio-economic and spatio-temporal perspectives that shape policy developments in the planning, development and proactive management of coastal environments. The paper identifies the need to accommodate a new civic awareness of potential environmental hazards in articulating the public interest and balancing the trade-offs involved.

Keywords: master planning; policy integration; public consultation; resort cycle; strategic regeneration

1. Introduction

The conditions and arrangements for the enhanced protection and management of the marine and coastal environments in the UK are matters of increasing concern (Department for Environment Food and Rural Affairs, 2002; Scottish Coastal Forum, 2004; Scottish Executive, 2005). At a human scale, Andrews et al. (2005), for example, report that 17 million people live within 17 miles of the England and Wales coast; that Britons took 23 million seaside holidays in 2003; and that 95 per cent of the UK's international freight passes through UK ports. Such statistics highlight the close relationship of many communities with the sea and indicate the strong economic and social interdependencies that exist between land, sea and community well-being. Yet, as the Marine Climate Change Impacts Partnership (2008), for instance, reports, climate change has a clear impact on the marine environment and ecosystem and increases the potential vulnerability of its associated goods and services. Indeed, evidence shows that average sea temperatures and sea levels have risen in the past century and that winter storms have increased, with the coast eroding at more than 25 per cent of monitored sites in England and Wales (Andrews et al., 2005). Moreover, scientific monitoring data relating to negative impacts on marine and coastal species and habitats reported by Andrews et al. (2005) suggest, in many cases, an increasingly uncertain future and that hard sea defences are critical to protecting coastal communities from increased flood risk. The consequences of beach erosion,

■ *Corresponding author. ***E-mail:*** **d.peel@ulster.ac.uk**

doi:10.3763/ehaz.2010.0048 © 2010 Earthscan ISSN: 1747-7891 (print), 1878-0059 (online) www.earthscan.co.uk/journals/ehaz earthscan

for example, potentially impact on a wide range of stakeholders, particularly those involved in tourism (Phillips and Jones, 2006). Yet, the nature of coastal intervention and operational developments in responding to these changing circumstances are dependent on a complex of state–market–civil relations and the incorporation of scientific evidence with respect to managed alignment of the coast (Myatt et al., 2003). This raises the importance of devising appropriate policy measures which can respond to environmental risks.

The management and regulation of the coastal–marine interface is rendered complex by its legal, institutional and property rights characteristics (McGlashan et al., 2004, 2005). In addition, the development of specific marine legislation has been identified as being further complicated by the devolved governance arrangements in the UK (Slater, 2007). In part, the challenges of orchestrating the strategic sustainable management of the coastal–marine interface and associated marine resources stem from the composite of private, public and common property rights involved (Feeny et al., 1996). Notably, the treatment of the seas as a 'free-for-all' has been attributed to the lack of a 'proper accountable forward planning system that enables us to protect the resources we rely on from irreversible damage' (Scottish Environment Link, 2006). A clarion call for a transparent, effective and integrated forward planning framework that acknowledges the interrelated nature of coastal and marine resources and a clear need for policy coupling (Connell and Flynn, 2003) is the starting point for this review paper. This builds on an acknowledged view that strategic responses to managing coastal environments necessitate an integrated and value-informed understanding of socio-economic, environmental science, technological and policy options (Turner, 2000). The discussion builds on an understanding of planned responses to risks of sea-level rise in terms of protection, accommodation or retreat (Pugh, 2004). It advances the case for an integrated response to the regulation and management of planned development that is sensitive to the socio-economic and environmental contexts and the vitality and viability of coastal communities.

The paper provides a complementary strand to the interest, for example, in the development of Shoreline Management Plans that similarly seek to overcome a previously piecemeal and relatively short-term approach to environmental management in this domain (Jennings, 2004). These parallel debates highlight a shift away from a mono-professional and engineering-focused approach to the coast to an integrated, co-ordinated and collaborative model of coastal management and defences (Potts, 1999). Moreover, the argument presented is consistent with a growing body of academic literature, which advocates the perceived synergies to be derived by advancing a natural and social science interdisciplinary appreciation of coastal management (Turner, 2000), promoting the management principles of Integrated Coastal Zone Management into practice (Phillips and Jones, 2006), and incorporating parallel relationships with the development of European spatial planning thinking (Kidd et al., 2003). The coupling perspective supports the view that coastal development and management requires a relatively more holistic understanding of the coast as a dynamic entity and part of a total-system framework that is time–space contingent (McFadden, 2008). Specifically, this paper takes a deliberate approach to reaffirm the strategic capacity of the land-use planning system to serve as the locomotive for regeneration, and to act as the decision space for stakeholder and community engagement over time.

In discussing the nature of contemporary coastal management, McFadden (2008, p. 311) asserted that, 'The impact of flooding and erosion, and how this varies in space and time, for example, depends on the adaptive capacity of the individuals, communities and organizations within these coastal environments. Yet we know relatively little regarding the spatial and temporal variability in the adaptive capacity of different coastal communities'. Deploying a case study approach, the paper critically discusses how an integrated planning strategy is being devised to secure the regeneration of the coastal resort of Porthcawl, South Wales. This seeks to show how a specific operating body and Coast Protection Authority, Bridgend County Borough Council (2002), is responding to the responsibility for assessing the need for flood and coastal sea defences and planning for reducing or managing that risk. At one level, a strategic spatial planning framework addresses a commitment to reinvigorate the local economy and enhance community well-being through the design of the town's seaside, promenade and harbour quarter as part of a concerted effort to improve the overall waterfront environment. Here, physical, community and economic priorities are articulated through relatively conventional strategic planning guidance. This planned strategy rests directly on the production of a master plan intended to secure a mixed use development scheme, which pays attention to the quality of urban design, strategic environmental assessment and active public consultation.

At another level, this land-use plan-led approach attempts to raise wider community awareness of the competing environmental scenarios associated with the town's coastal location. On the one hand, climatic risks potentially threaten the resort in terms of sea-level rise, increased frequency and intensity of storms and coastal erosion. Flooding potentially threatens the community in a number of ways in terms of risks to homes, businesses, services, infrastructure, livelihoods and reputation. On the other hand, the decline of the resort is also threatening the viability and vitality of this coastal town. The proposed regeneration strategy explicitly seeks to enhance the town's attractions and revitalize its tourism potential. Moreover, flood defences and beach management initiatives form an integral part of a strategic developmental framework and implementation plan which seeks to anticipate the impacts of climate change. In providing a historical perspective to the settlement's development, this paper illustrates how a small town has adapted over time. This serves to explain the role that the land-use planning system has to play in improving the resilience of coastal communities to respond to socio-economic decline and the risk of environmental hazards.

The remainder of the paper is structured as follows. First, it provides a background to planning and regeneration in order to contextualize the emergent thinking around the need for integrated policy approaches to the management of change and improved responsiveness to climatic risks. This is followed by a discussion of the resort cycle model, which helps to explain the external and internal conditions and prerequisites of seaside resort restructuring (Agarwal, 2002). Taken together, these sections advance a conceptual model for understanding planning and regeneration in the coastal environment within a prevailing growth paradigm that emphasizes economic development. The main sections elaborate the rationale for, and operationalization of, land-use planning in the UK. Reflecting a growing concern within the UK planning profession to address the challenge of climate change (Royal Town Planning Institute, 2009), the example illustrates the extent to which policy responses to environmental hazards in the coastal context may be framed by a wider concern with the integration of scientific evidence and evidence-based policy action, stakeholder engagement and social learning.

2. Planning for sustainable coastal communities

There is an emerging policy interest in the economic, social and environmental well-being of coastal towns in the UK, which reveal a growing awareness that they are increasingly exposed to complex processes of economic restructuring, community changes and environmental vulnerabilities (House of Commons Communities and Local Government Committee, 2007). For many smaller coastal towns there is evidence, for example, of relative underperformance in terms of employment, property development, the provision of community facilities and investment in the public realm. A number of coastal locations are increasingly subject to unemployment and a general deterioration in their physical environments. In terms of the public realm, for instance, facilities and retail services in many coastal towns are becoming threadbare as their economic base becomes more fragile. These restructuring processes and outcomes will likely be articulated in different ways and over different time scales, reflecting the particular conditions, locations, histories, experiences and the character and effectiveness of any state responses put in place, but a dominant contemporary concern relates to the need for appropriate policy intervention to regenerate and sustain coastal communities and to arrest their decline.

A powerful countervailing reasoning asserts, however, that coastal towns have the potential to create new identities with a greater potential for tourism, sporting and other related economic opportunities (English Heritage and Commission for Architecture and the Built Environment, 2003). Such arguments make a number of assumptions about the character of the new world in which coastal towns can reconstruct themselves. The capacity of small towns for regeneration in functional and strategic ways is a critical prerequisite to these ambitions. For example, attention must be paid to the complexities of global economic conditions, international competitiveness, the variability of exchange rates, individual perceptions of such coastal locales and the capacity of local leadership. These different perspectives are not of course mutually exclusive to the wider concerns of regeneration and the promotion of sustainable communities. Moreover, these debates must also take account of the specific environmental contexts and the need to incorporate fresh evidence into the strategic vision for an area. Working within a multi-professional context requires an integrated understanding of the issues involved.

A number of contextual points are considered relevant to understanding regeneration and its application in the coastal resort context. First, regeneration tends to be an established state activity that seeks to address the processes contributing to urban decline, and to promote the conditions which are held to be appropriate to securing physical and socio-economic renaissance. Regeneration is principally associated with managing the post-industrial conditions of Britain's established manufacturing cities (Carley, 2000). In general terms, this policy sphere has been designed specifically for relatively larger urban localities and has tended to be organized around specific state-based area and institutional interventions. These have involved various combinations of community measures, planning strategies and property development initiatives to secure desired policy outcomes (McCarthy, 2007). Evidence shows that there is an established tradition and experience of regeneration involving a range and combination of possible social, economic, environmental and cultural imperatives (MacLeod and Goodwin, 1999). Yet, regeneration in the UK is not uncontested. There are a number of fundamental concerns regarding the design, resourcing and effectiveness of different approaches to urban regeneration (Gripaios, 2002). In particular, in this context, attention has been drawn to the extent to which regeneration has involved active interventionism or simply seeks to mimic market actions (Fuller and Geddes, 2008; Boyle et al., 2008). Notwithstanding these broader practical, institutional and ideological questions, in recent times, political and policy interest has moved to a concern with regenerating coastal settlements primarily in order to optimize their economic and social potential.

Second, an understanding of the role and design of the necessary institutional arrangements involved in regeneration is important. While public–private partnerships have been an enduring feature of local regeneration (Lawless, 1994; Fernie and McCarthy, 2001), in practice, regeneration delivery vehicles have evolved to comprise various combinations of public sector and private resources and to rest on different models of partnership working (Whitehead, 2007). Moreover, urban regeneration practices have matured so as to engage at different scales – regional, local and site-specific or development project levels (Fuller and Geddes, 2008). This reflects a concern with matching spatial priorities with the appropriate jurisdictional and institutional agencies involved. More recently, this catalogue of potential measures has been supplemented with an interest in promoting active competitiveness, devising alternative investment vehicles, providing training tools and facilitating institutional mechanisms at different scales so as to construct a new sense of place. This is grounded in a commitment to community engagement in neighbourhood renewal and the promotion of social learning for developing sustainable communities (Office of the Deputy Prime Minister, 2002). Critically, this involves a concern with promoting developments that can encourage behavioural change to secure relatively more sustainable futures.

Third, a new interest concerns the role of promoting a greater sense of identity representation in local development and redevelopment (Lloyd and Peel, 2008). This is important in technocratic or operational and capacity terms as well as the democratic aspects of gaining wider social acceptance and political legitimacy for the economic transformation of particular resorts. While the main emphasis of identity representation thus far has tended to be drawn from the experiences of older, larger industrial cities, identity will be of considerable importance to the well-being of smaller coastal towns. Here reference has been made, for example, to the importance of enabling effective brand management in, and of, small towns through the deliberate linking of marketing principles to their specific locations in an attempt to remain competitive or as the basis of a turnaround strategy in response to economic and social decline (Kerr and Johnson, 2005). Reputation may then be presented as critically important to successful regeneration.

Different circumstances and influences therefore combine to create a deliberate interest in devising appropriate redevelopment responses in coastal towns. Critically, it is the particular coastal environment and attendant hazards which demand that policy makers also take account of emergent evidence in relation to, for example, the increasing frequency or intensity of storms and the risk of surges, which necessitate an appropriate planning response in terms of location, siting, design, layout and infrastructure. The case study discussed initially draws on the life cycle analytical framework associated with the evolution of tourism destinations in order to provide a conceptual framework for tracing the evolution of a specific coastal town, Porthcawl. This model provides the background to explaining conventional understandings of resort management and development.

3. A resort cycle perspective

The resort cycle model of seaside resort change initially developed by Butler (1980) provides one explanation of the adaptive capacity of traditional British seaside resorts (Agarwal, 1997, 2002). This particular social construction of change stresses a linear progressive sequence in which resorts are progressively explored, established and consolidated, and then prosper and potentially decline over time as a consequence of, inter alia, increased competition from overseas resorts and specialist activity-based vacations elsewhere (Gale, 2005). Here, the implementation of environmental management policies is critical in seeking to address associated issues (Jennings, 2004). It is argued that as the established facilities of a resort increasingly exhibit the effects of accumulated wear and tear, places tend to become less competitive. This then inhibits their vitality and viability in sustaining their well-being.

In general, the resort cycle model of seaside resort change rests on an iterative number of phases, suggesting that stagnation and decline may be addressed through the provision of deliberate counter-measures. In this context, an active role for regeneration is held to be important (Agarwal, 1997). As a resort experiences a relative downturn in its fortunes so a managed intervention to address the associated effects may be necessary. Furthermore, such action can then explore ways in which the decline can be challenged, reversed and/or redirected. Indeed, there is an established history of coastal resort regeneration through the design and implementation of local economic development initiatives involving what may be considered to be variants of more mainstream urban regeneration measures. These need to be highly sensitive to a range of internal and external explanatory influences of decline and wider structuring opportunities for change (Agarwal, 2002). Furthermore, there is increasing evidence that these interventions must be layered on an appreciation of the wider potential impacts of sea-level rise and erosion, for example, as a new set of environmental vulnerabilities emerge.

There are a number of criticisms of the resort cycle approach and its conceptual basis has been recast to question the explanation of resort decline as a life cycle process and to promote the need for a greater appreciation of place distinctiveness in resort cycle restructuring (Agarwal, 2002). As a general model, it cannot capture the very different specificities of individual locales; in effect, there is no one standard coastal resource – each reflects a particular combination of dimensions, including location, geomorphology, culture and socio-economic development. Thus, a contrast may be drawn, for example, between beach resorts that fall within a city region, fishing and marina centres that form part of an actively marketed network of facilities, or relatively more remote centres for marine ecotourism. Alternative understandings of coastal settlements offer the potential to promote a deeper structuralist analysis of economic restructuring and cultural change, which impact on traditional seaside resorts (Gale, 2005). A relatively greater sensitivity to climate change and coastal erosion and flooding, for example, also needs to be accommodated. Here, a parallel may be drawn with Holling (2001) with respect to the complexities involved in cycles of managed adaptation. This logic suggests the need to be sensitive to the particular hierarchical relations involved in the various scalar, economic, ecological and social systems of coastal resorts and how policies and interventions are designed, prioritized and managed in practice.

The vulnerability of coastal resorts not only stems from global economic changes and shifting tourism preferences, but the scientific evidence relating to specific coastal hazards due to erosion, storm surges and risk of flooding. Bearing in mind these caveats of the resort model, and combining these insights with those from the regeneration literature, Figure 1 provides a conceptual basis for tracing change and projecting regeneration in a coastal town. This modifies the resort timeline by presenting a model that identifies a plural commitment to regeneration that transects the decline curve but then offers an idealized model of sustainable regeneration that presents a strategic approach. Moreover, a commitment to intervention to address the perceived features of decline are complemented by a responsiveness to local community engagement in envisioning a future for the locality, based on a concern with fostering an informed civic response to the management of change and stewardship of environmental resources (Peel and Lloyd, 2007). In practical terms, a strategic planning approach to regeneration may be secured through appropriate governance arrangements that are sensitive to prevailing and anticipated economic, environmental and social interests.

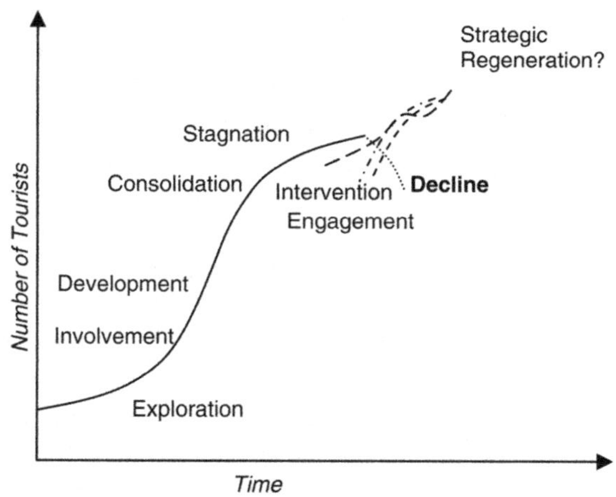

FIGURE 1 A strategic planning approach to resort regeneration

Source: Jennings, 2004.

4. Planning for regeneration

This section concerns the sphere of regulatory arrangements associated with land-use planning decision making that have the capacity to create the necessary political space to deliberate and secure the regeneration of coastal towns. Political space is held to be an important strategic dimension to the more conventional activities around state regulation, visioning and development (Institute for Public Policy Research, 2008). Here, the role of the statutory land-use planning system in seeking to enable the creation of political space is effectively a deliberate assertion to locate issues associated with coastal towns in a wider socio-economic, environmental and spatio-temporal discourse. In effect, the system is attempting to broaden the political understanding of the complex nature of regeneration, which is not necessarily confined to the older manufacturing urban sphere. Smaller towns, including coastal resorts, are not immune from deep-seated processes of economic, industrial and social restructuring, nor the broader spatial and ecological effects which may represent a second phase impact on the coast as a consequence of climate change. The creation of political space seeks to promote a broader understanding of the relative conditions of smaller coastal towns, and place these in the context of emerging scientific evidence in relation to sea-level rise, potential flooding, erosion and new forms of policy response. In the context of plan-led regeneration, planning can serve to enable the delineation of the necessary democratic space through civil engagement to inform the generation of an evidence-base to guide the decision making, management and operational nature of development. Moreover, planning has the potential to offer a strategic prism to secure the intended regeneration outcomes, and provide the management and enforcement regimes necessary to monitor and regulate environmental and developmental impacts over the longer term. Importantly, high-quality design is identified as integral to the planning process and is held to be 'indivisible from good planning' (Office of the Deputy Prime Minister, 2004, p. 14). This is an important policy ambition.

In the UK, the use and development of land is controlled and regulated principally through statutory processes of devolved local decision making. The Town and Country Planning Act 1947 nationalized the rights to future land and property development and distinguished between the physical development of land and material changes in its use (Booth, 2002). In practice, the UK planning system is characterized by a culture of discretionary decision making that requires that individual decisions are made on a case-by-case basis, which takes account of the relevant material considerations (Booth, 1995). This stresses the local focus of the planning system in that it has the potential to craft development schemes that reflect prevailing conditions and changes in specific localities. An integral aspect of the planning system is that the local focus takes place within a strategic framework that reflects national policy priorities and demands for infrastructure, and allows for securing an integrated approach to territorial management through community engagement and stakeholder consultation.

Importantly, planning has an acknowledged history and experience of public participation that provides a wider civil and political legitimacy (Davies, 2001). While the anticipated outcomes of what is held to be the public interest are contested, civil engagement remains an intellectual driver for planning action. Over time, planning has evolved and responded to changing conditions, contexts and priorities (Poxon, 2000). In this way, planning has attempted to be sensitive to emerging social, economic and environmental issues, such as sustainable development, social inclusion and regeneration (Healey and Shaw, 1993). In practical and operational terms, planning decision making in the UK has followed a plan-led principle since the 1990s and there is a statutory duty to prepare a development plan to guide future development and promote certainty and consistency to investors and developers. Importantly, this forward-planning activity places a clear emphasis on monitoring and evidence gathering as the basis for effective policy making.

Notwithstanding a concern that land-use planning has, to date, taken insufficient responsibility for flood and coastal management, a recent development has been a greater sensitivity to environmental hazards, vulnerabilities and sensitivities, and involved enhanced guidance for the management of this resource (Howe and White, 2004). Specifically, debates with respect to the rationale for mitigation and/or adaptation are being mainstreamed within the statutory land-use planning system. The underlying premise is that there is a need for both political legitimacy and community support for the proposed development – whether it be private sector or government-led in origin. Importantly, development plan frameworks explicitly set out the environmental constraints within which development can or cannot take place. Critically, there is an asserted and deliberate interest in securing sustainable development that takes appropriate account of the emerging scientific evidence in relation to climate change.

In making decisions about using and developing coastal areas, development plans are the primary material consideration in the statutory decision-making processes of the UK planning system. These various plans draw together the relevant strategic and local policy frameworks for determining individual land and property development proposals. These formal arrangements allow for international, national, regional and local policy considerations to be taken explicitly into account with, in the majority of cases, decisions then being taken at the local – and therefore site-specific – level. In effect, the technocratic requirements of policy implementation are complemented by the democratic underpinnings of the discretionary planning system. The forward planning approach to the management of the land and coastal resource has directly informed emerging debates around the case for, and potential of, marine spatial planning (Peel and Lloyd, 2004) and the associated regulation and management of competing interests in the coastal zone. Most recently, this has witnessed an extension of the meaning of development to include schemes located in the marine environment (Peel and Lloyd, 2008). This social reconstruction of the coastal resource and understandings of the potential of the marine demands a new political space for debate about the risks and potentials involved in development.

In order to reconcile economic, social, scientific and environmental priorities that may be variously driven by political, government, community, campaign group and business agendas, there is a need to facilitate what Connell and Flynn (2003) describe as policy coupling. Here, the strategic imperative is consistent with the principles and outcomes of sustainable development. In effect, the statutory planning system can serve as a conduit to identify the social, economic and environmental dimensions to a problem, and then act as a way of mediating to secure an outcome consistent with what is held to be in the public interest. This process of reconciliation of pressing community-based concerns and economic ambitions is clearly tempered by environmental limits. Balancing these tensions and legitimizing affirmative action is then the principal focus of planning as a regulating and regulatory tool. In practice, generic planning principles are specifically designed to be shaped by local communities and economic, environmental and social conditions. A case study of Porthcawl, South Wales, a relatively small coastal resort, is examined to illustrate a planned approach to regeneration, which is also responsive to local concerns about climate change.

5. Porthcawl: Towards a regenerative resort cycle perspective

Devolution has created a new territorial politics in the UK (Jeffery, 2007). While retaining the generic function and form of planning, political devolution has allowed for the crafting of more distinctive regional approaches to land-use planning. In understanding the implications of coastal development in the context of Porthcawl, it is important to note that the National Assembly of Wales has inherited many of the duties, powers and responsibilities that were formerly held by the Secretary of State for Wales. While the National Assembly has no powers to make or amend primary legislation, it can make and amend policy and secondary legislation, which is of significance to the implementation of the land-use planning system and regeneration. It operates within legislation designed for England and Wales, although Section 6 of the Planning and Compulsory Purchase Act 2004 provides the opportunity for the Welsh Assembly Government to devise land-use planning arrangements, which are considered more appropriate to its territorial needs. The legislative provision has allowed the Welsh Assembly Government to innovate with respect to the design of a spatial planning approach for the nation-region and to devise appropriate local planning measures for territory and developmental conditions.

Indeed, in comparison with other devolved territories, Wales is held to have developed a number of novel policy agendas (Bradbury, 2005), including a National Spatial Planning Framework (Powell, 2001). There was an early recognition of the need for a strategic planning framework for the Welsh economy, which was intended to secure wider integration of agencies, sectors and local planning (Harris et al., 2002). Critically, this was complemented by an emphasis on fuzzy boundaries and effective co-operation and co-ordination between public sector agencies (Welsh Assembly Government, 2004). With respect to the coast, there is an emphasis on coastal management in such a way as to link spatial planning on land with spatial planning at sea, in line with the objectives of the Integrated Coastal Zone Management Strategy for Wales. In strategic and institutional spatial planning terms, the Welsh Spatial Plan is principally concerned with devising appropriate governance arrangements for cross-sector working (Harris and Hooper, 2004). Significantly, the Welsh Coastal Tourism Strategy (Welsh Assembly Government, 2007) was identified as one of the key national actions within the Wales Spatial Plan in order to nurture the sector's economic potential.

In spatial representational terms, Porthcawl features as an interdependent settlement of a more extensive Swansea Bay Waterfront and Western Valleys spatial plan area that seeks to create a modern and well-connected city-region characterized by ambitions for a competitive, knowledge-based economy, high quality of life, sustainable environment and a vibrant waterfront (Welsh Assembly Government, 2008). Within this broad strategic context local planning and regeneration measures for localities, such as Porthcawl, are then devised and executed. Devolution and governance reform, policy design and partnership working alongside assertive planning represent a new strategic prism in which this particular resort is being recast.

The earlier development of industrial towns on the western edge of the south Wales coalfield, such as Llanelli, Swansea, Neath and Port Talbot, was based, to a large extent, on the growth of metal industries (Newell and Watts, 1996). Towards the end of the 18th century, a rich coal seam was discovered in Maesteg, South Wales. The subsequent need to transport the coal by ship led a number of entrepreneurs to search for a suitable coastal outlet. Porthcawl was identified as a potential port as its location was in relatively close proximity to the coal fields. This potential led to an Act of Parliament for the construction of a tramroad from Maesteg to a new dock at Porthcawl.

Porthcawl and its harbour played a strategic role in the broader development of the South Wales coal economy. The small harbour did not prove entirely satisfactory, however, and was subsequently enlarged in 1840. The demand for coal and then iron required the infrastructure to be improved yet again, and a new railway company was subsequently formed to lay a broad-gauge steam line from Maesteg, via Tondu and Cefn, to Porthcawl. Private investment, industrial expertise and a skilled workforce all underpinned the early strategic industrial importance of the town (Higgins, 1968). Notwithstanding the expanded coal export trade from the extended harbour, towards the end of the 19th century Porthcawl was unable to compete effectively with the changing requirements of iron processing, the increasing size in shipping and the competition offered by the relatively larger and more modern docks at Port Talbot (to the west) and Barry (to the east). Furthermore, Porthcawl's potential expansion was limited by the nature of its tidal docking and its relatively high exposure to marine and coastal elements. Indeed, the proximity of sandbanks and a nearby treacherous reef, Tuskar Rocks, explain, in part, the significant number of wrecks in the area; some 300 were recorded in the 19th century alone (Morgan, 1975). Indeed, it was this specific environmental hazard that impeded the continuing industrial development of the town at this time.

Nevertheless, the earlier works involved in constructing the tramroad and the port activities had increased the demand for housing and related developments, such as hotels and public houses. These proved to be an important legacy of the early industrial period of the town. What was described as haphazard development serving the industrial needs of the area continued until 1864 (Morgan, 1975), however, when an important turning point was reached in the town's history. An established local entrepreneur and landowner, James Brogden, had the vision of Porthcawl as a seaside town with a promenade, hotels and elegant houses. This represents an early attempt at resort town planning and it attracted others to become involved in its development. These visionaries included Dr James Lewis and Florence Nightingale, who founded the 'Rest' Convalescent Home in 1862. The pace and scale of the resort's subsequent development resulted in the formation of Porthcawl Urban District Council in 1893. Attempts to protect the nascent tourism asset and hold back the tide resulted in the construction of the Esplanade sea wall in 1877, which subsequently had to be replaced in 1906 and 1934. Phillips and Jones (2006) not only highlight the expenditure involved in erecting such engineering structures, but also the potential erosion that such methods can themselves promote. Indeed, they further note the less than aesthetic paving of part of the beach at Porthcawl with bitumen macadam in 1984 as an alternative mode of protection as beach levels continued to fall. This revetment was specifically provided in order to prevent erosion and scouring of the sea wall, overtopping of the promenade and the flooding of properties in the vicinity.

Despite the decline of the industrial port, Porthcawl's transformation into a small coastal resort was becoming more evident. Its potential as a bathing venue and location for day-trippers, for example, many of whom travelled down from the South Wales valleys by rail, was increasingly recognized. The established railway service also made Porthcawl attractive to businessmen from Cardiff, who established an early commuting community. As a result, Porthcawl grew steadily, both as a seaside resort and, increasingly, as a coastal residential town, until the outbreak of war in 1914. Between the wars, Porthcawl further developed as a holiday resort and continued to expand, particularly as a consequence of a building subsidy provided for the depressed South Wales regional economy. Increasingly, there was greater involvement of local people and the enhanced provision of services and an extension of the town's market area. Porthcawl's prospects as a resort continued with the development of a number of attractions. A Pavilion and Winter Gardens were opened in 1932 and a lower promenade opened in 1935. The remains of an allotment were redesigned as public open space – Griffin Park – with a range of leisure facilities, including bowling and putting greens and tennis courts. These facilities provided a new aspect of, and access to, the town's coastal location and natural assets (David, 2006). Porthcawl thus developed as a resort, thereby replacing its initial role as an industrial harbour.

The consolidation phase of Porthcawl as a coastal resort principally centred on the development of its tourist attractions. An old ballast-tip had already been used as the site of a 'figure eight' wooden roller coaster and other amusements since the early 1920s. Porthcawl Recreations Limited was formed in 1924 to develop the site into an amusement park that included a water chute, big wheel and world cruise and dodgems roundabouts. As a consequence, Coney Beach Amusement Park (established in 1939) became a widely known leisure venue in South Wales. In 1945, the Eastern Promenade Development Scheme was completed, including the entrance to Coney Beach and what was known as the 'Cosy Corner'. This development scheme extended the seafront promenade and transformed a previously derelict site, itself a legacy of the closure of the dock, into an attractive public space. Camping had increasingly become popular during the interwar years, so much so that by 1939 there was a temporary influx of 12,000 campers over the bank holiday weekend. Later, canvas was replaced by trailer-caravans and temporary tents became permanent holiday camps. Higgins (1968) notes how by 1950 there were 11 such sites in the parish, including Trecco Bay, another popular landmark in the resort.

Following the reasoning of the resort cycle model (Figure 1), the (inevitable) stagnation of Porthcawl subsequently followed these periods of establishment, potential and growth. As with many other resorts of this type, the 1950s began to witness very evident changes in domestic holiday patterns. This resulted in a shift away from family holidays to day-trippers or coach parties, particularly to the Coney Beach Amusement Park and Sandy Bay. This prompted a closure of many of the boarding and private houses that had previously taken regular family visitors over the summer. Over time, the fortunes of Porthcawl waned further, and the complex economic and demographic restructuring that has taken place has made it highly vulnerable in terms of its sustainable viability and vitality. Similar to other seaside resorts in South Wales, Porthcawl began to suffer increasingly from a negative image (Selby and Morgan, 1996). Slowly, it began to illustrate the signs of a deteriorating town even though its popularity as a commuter town continued, especially for Cardiff. Yet, the town itself was in need of deliberate attention. In effect, a relatively long period of stability was succeeded by one of gradual instability that gathered its own momentum. The failure of the business sector in the town to adapt to and counter this process through new investment resulted in the continuing relative deterioration of the resort. Global changes and pressure served to exacerbate its economic vulnerability, and increasing environmental awareness of the potentially negative impacts of climate change necessitated a new integrated strategy to sustain the town's socio-economic and cultural viability and vitality. This fell to the local authority to lead and manage within the context of devolution.

6. Post stagnation and Prosiect Saith Bae – the Seven Bays Project

In 2004, Supplementary Planning Guidance to the Bridgend Unitary Development Plan, the Porthcawl Development Framework, was adopted by the Borough Council, covering the Porthcawl Regeneration Area (Bridgend County Borough Council, 2004). The comprehensive Porthcawl Waterfront regeneration area encompasses the harbour, Cosy Corner, the fairground site and Trecco Bay. The regeneration of the foreshore and former dockland environments in the resort was asserted as a potential vehicle to providing an important new focal point to create social, economic and environmental benefits for the town itself and the wider area. The scale of the regeneration site, its southerly outlook, its expansive views of the Bristol Channel and the English coast beyond, and its relationship with the existing town morphology were perceived as creating unique features – not only in the context of Welsh seaside towns, but also within the UK.

Supplementary Planning Guidance (Bridgend County Borough Council, 2004) was produced to promote Porthcawl and its waterfront as a major regional tourism asset; to reinforce the positive identity and attractiveness of the town, to provide a flexible strategic planning framework for public and private investment and to establish a consensus for change. Owing to a number of changing issues, circumstances and thinking, a subsequent version of the Waterfront Planning Guidance – Seven Bays Project – was published in 2007. This asserted that the

> regeneration of the waterfront is a key element in the future development of Porthcawl as a premier seaside resort. This will be achieved through a combination of quality design of buildings and the surrounding environment, comprising residential, leisure, cultural, tourism and retail uses. Central to this will be a revitalised harbour area incorporating the reuse of historic buildings, new leisure facilities and a fully refurbished harbour set within a prestigious and vibrant waterfront (Bridgend County Borough Council, 2007a, p. 3).

The revised strategic planning guidance was the subject of an extensive six-week public consultation (Bridgend County Borough Council, 2007b). This included a newsletter, public exhibitions, letters to statutory and other consultees, media and website coverage and a number of public meetings. The clear majority of respondents acknowledged the need for change – particularly with respect to the proposed seafront promenades, the harbour quarter, enhanced leisure facilities and the beach environment – and broadly welcomed these as fundamental to Porthcawl's anticipated new civic realm. This participatory content enabled the modification of the planning guidance to reflect local concerns and to take into account new scientific evidence with respect to climate change. This extended appreciation of the material considerations to be taken into account is critical since the increased capital costs for sea defences were integrated into developer planning obligations (Bridgend County Borough Council, 2007c). This decision appears to respond to the concerns set out in the Shoreline Management Plan (Bridgend County Borough Council, undated) regarding the complex ownership and poor physical conditions of the sea defences in Porthcawl. Moreover, a sustainability appraisal advocated the need for further studies and continuing scrutiny of potential flooding risk, while stressing that conventional technical solutions would likely be appropriate in such circumstances (King, 2004). The holistic and iterative nature of this planned approach represents a contemporary precautionary attempt to implement strategic regeneration through planning for sustainable futures in the light of climate change.

7. Conclusions

In this paper, and developing the broad logic of the resort cycle model (Figure 1), a case study of Porthcawl has been used to illustrate the adaptive potential of a coastal town resort over time. The particular coastal environmental context examined served both to expand and reduce the development potential of the settlement and may be headlined in a number of broad iterative developmental stages. These then illustrate how planning, as an evolving activity, served to promote adaptive capacity. In recent times, planning itself has adapted to address regeneration activities directly and to accommodate emergent evidence and understandings of the potential consequences of climatic change. This case study then responds to McFadden's (2008) exhortation to develop a broader understanding of societal, community and institutional responses to the management of change and the stewardship of built and natural environments.

From industrial beginnings, Porthcawl morphed into a resort because of its inherited infrastructure, proximity to Cardiff and Swansea, and its coastal location and harbour. In many ways, an exemplar of a small UK coastal resort, Porthcawl boomed and declined. Its deterioration was endemic and took place over a number of decades. Devolution, a refreshing of locally sensitive policy agendas, and the deliberate use of a reformed planning system then served as a mechanism for securing the longer term sustainability of the town and its regeneration. The Seven Bays project seeks to revive and replace a core element of the resort – its waterfront and historic harbour – while responding to fresh evidence about the potential risks of environmental hazards, and the needs of the wider community. The replacement of a declining seaside sector with a new mixed-use development scheme also recognizes the opportunity to manage potential coastal hazards, particularly erosion and flooding, as part of a holistic economic, social and technical response to local conditions. Here, however, the scientific evidence remains partial.

The revised master plan (Bridgend County Borough Council, 2007a) changed the balance of the initial mixed-use development scheme by increasing the number of residential units, and proposing a larger food store to support the town. In addition, it no longer included a marina on financial and capacity grounds, but instead proposed other leisure uses around the refurbished harbour area. Significantly, flood defences were put forward to address specific concerns relating to climate change in the community consultation (Bridgend County Borough Council, 2007a). This explicit recognition of the need to incorporate environmental hazards into the conventional metrics of regeneration is significant. It reflects the regional perspective of the Welsh Assembly Government in explicitly seeking to reconcile economic, environmental and social ambitions, and a local concern by the Council to mitigate change and to reduce future risks. Pragmatically, this approach requires incorporating and integrating, where appropriate, the available options and financial means to secure the necessary capital costs for flood defences. The implications of this re-prioritization of policy imperatives diverts expenditure from elsewhere and requires an appropriate decision-making context for securing consensus; the statutory planning system provides that democratic political space.

It is clear that climatic change and the potential risk of more frequent and more intense storms, for example, demands a higher profile of environmental hazards in discussions and decisions about the nature of planned human development. While the planning system provides a tested and legitimate arena for marshalling evidence, considering and weighing up material issues, and deliberating options with respect to longer-term visions for sustainable settlements, new understandings of the implications of climate change necessitate an explicit incorporation of the available evidence in clearly articulated ways. Forward-looking development plans have to be cognisant of the highly differentiated conditions and potential socio-economic, cultural and ecological vulnerabilities of individual coastal towns. Conventional concerns of the planning system are now required to anticipate potential environmental hazards which will vary over time. Yet, this end point is elusive and the scientific evidence is uncertain (Parliamentary Office of Science and Technology, 2009). This then demands appropriate ways of debating and planning for risk. New thinking is thus required if planning is to maintain a proactive approach to the integrated management of change and to sustain community well-being at the regional scale.

Importantly, attention has been drawn to the need to tighten the regulatory activities of planning in order to take into account coastal erosion and risk (O'Riordan et al., 2008). This enhanced awareness is evident in the incorporation of climate change within the educational framework of the UK's planning profession (Royal Town Planning Institute, 2010). Yet, while the importance of planning as a statutory regulatory process in securing long-term societal ambitions is asserted, there is also a tension associated with the normative prescription of what planning *should* do. Planning is a tool of the state, and, as such, seeks to deliver the public interest. Priorities and interests vary, however, and society has to define the public interest over the short, medium and longer term. As the case study illustrates, coastal erosion and flooding need to be seen on a broader canvas of socio-economic regeneration and strategic regional viability and not simply reduced to a local planning or technical matter. Trade-offs have to be made, and public engagement in understanding and debating the options is a prerequisite if communities are to understand the rationale for and against specific development proposals.

In line with the adopted policies with respect to conservation, coastal and shoreline management, and tourism development, the strategic master plan for Porthcawl seeks to articulate a balanced view, which recognizes that environmental vulnerabilities have to be incorporated within broader socio-economic and community regeneration thinking. The statutory consultation processes afforded by the land-use planning system enabled the authority to include environmental considerations into its regeneration agenda for this small resort. A significant feature of this case study is the deliberate attempt to integrate and off-set the costs of addressing the potential environmental hazard of coastal flooding and erosion through private developer contributions. Here the state – Bridgend County Borough Council – is proactively ensuring the mitigation and adaptation to a potential environmental hazard by clearly identifying the necessary infrastructure required to support the new development. In the case study, a strategic planning dimension allows for the reconciliation of competing ambitions, the physical transformation of a declining resort, with sensitivity to the perceived environmental realities. As such, this redevelopment master plan approach directly addresses the governance and physical conditions of Porthcawl. It specifies the necessary public–private institutional roles necessary to the sustainable regeneration of the resort, and casts the regeneration project into the functional resort region, reflecting a strategic concern with the seven bays and wider spatial plan for the Swansea city-region.

The overall concept for the longer term viability of this established settlement and the regeneration of Porthcawl rests on the current vogue for a mixed-use resort development approach (Barnett, 2005). Within the commercial land and property development sector the potential of a mixed-use scheme is acknowledged as generating considerable advantages to developer, occupier and consumer (Parker and Garnell, 2006). In practice, this approach can spread the risks attendant on such investment strategies. The Porthcawl example shows how the nature of this seaside resort is not simply being replicated but is being diversified to attract different markets, all with various interests and motives. Moreover, the regeneration strategy seeks to integrate responsible environmental management as part of the initiative (Walker, 2008). This policy coupling approach is reflected in Figure 1 which indicates that there is no single strategy but a strategic response to the recognition of decline.

In this context, the statutory land-use planning system, through the preparation of supplementary planning guidance, asserts a master plan approach to facilitate a coherent and consistent planning strategy intended to lever in the required private sector investment to protect the public interest. The significance of bringing together economic, regeneration and social facilities with appropriate sea defence and flooding infrastructure provision demonstrates the progressive coupling of policies and the associated institutional resources that orchestrate private and public actions. The political weighting of the priorities inherent in the planning process is evident in the policy iterations and emphasis in the context of changing circumstances, public and stakeholder engagement, and, significantly, in the light of a greater recognition of the threats of climate change.

In mobilizing understanding and support for change, the example discussed illustrates an attempt to provide a strategic framework to the regeneration of the waterfront and the town with civil engagement and involvement through deliberate consultation processes and an explicit attention to climate change issues. A planned approach is not without its difficulties. Research evidence elsewhere has suggested that the dynamics of the participation arrangements put in place can themselves serve as a barrier to the effective input of local knowledge (Fudge, 2007). The case study revealed differentiated community interpretations of planning and regeneration in Porthcawl from the assertion of the imperative for action, support for the plans, disappointment with the limitations of the proposals and inadequacy of the provision of community facilities, and reference to past deficits in local and regional governance (Bridgend County Borough Council, 2007a, b). A lack of earlier investment was identified as having reduced the resort's capacity to adapt. A combination of devolution, research evidence around the vulnerablity of UK coastal towns and climatic change, and the preparation of a new tourism strategy for Wales highlighted the latent potential of Porthcawl in a strategic capacity. This provided the political impetus for a local development framework to support the town's regeneration. The perceived need for flood defences added an important financial dimension that required trade-offs to be made with respect to local community expectations.

The Porthcawl case study provides a contemporary attempt to acknowledge and integrate an economic–ecological relationship in a local regeneration strategic framework that then nests within national–regional spatial priorities. This logic follows the reasoning of Connell and Flynn (2003) who advocate such policy coupling to secure more effective sustainable development outcomes. In this case, the linking of the sea defences as an intrinsic component of the strategic priorities of the regeneration plan directly addresses the location and vulnerability of the town and its intended new land and property development investments. Nevertheless, the linking of the environment control elements of the master plan with the wider regeneration aspects may be interpreted primarily as an instrumental rather than a transformative attempt to accommodate economic and environment values with climatic risks. Essentially, the regeneration strategy stresses the importance of a design-led, inclusive, collaborative, focused and master planned vision to address the resort's perceived stagnation.

Taken together, these issues raise significant questions for the role of the planning system in orchestrating appropriate managed responses. Extreme climatic conditions necessarily require a positive land-use planning framework to offset potential crises and anticipate appropriate recovery programmes (Pitt, 2008). The implications are that appropriate policy responses to climate change are required to be identified, debated and planned for so that the integration of technological interventions in development schemes can be secured (Stern, 2007). The current interest in facilitating economic recovery through the deliberate investment in green infrastructure, including flood protection and coastal defence schemes, underlines the potential of the planning system to act as a strategic mediating agency (Bowen et al., 2009). Land-use planning can therefore be seen as assuming a more integrative role in forward planning, public policy design and intervention for regeneration in the short, medium and longer terms.

Through its strategic forward planning functions and tactical development management activities, the case study of Porthcawl provides a contemporary example of how the statutory planning system provides an important cypher for the collation, interpretation and integration of emergent scientific evidence and the mediation of tentative and uncertain understandings of environmental hazards and climatic risks as part of a proactive response. The experience suggests that balancing local community amenity concerns and the socio-economic imperatives of national governments against the perceptions of environmental risk will ultimately depend on the plausibility of the scientific evidence, the perceived validity of alternative governance scenarios and the legitimation of the developmental time frame involved. It confirms that placing future development scenarios within the context of socio-economic change alone is insufficient and indicates that strategic restructuring has to take specific account of uncertain environmental conditions and the potentially costly implications of unpredictable environmental hazards. Strategic planning provides a legitimate and fundamentally political space for identifying and weighting inter-scalar and cross-temporal policy priorities in the light of socio-historical constructions, scientific projections and competing alternative economic futures.

References

Agarwal, S., 1997. The resort cycle and seaside tourism: an assessment of its applicability and validity. *Tourism Management*, 18. 65–73.

Agarwal, S., 2002. Restructuring seaside tourism – the resort lifecycle. *Annals of Tourism Research*, 29. 25–55.

Andrews, J., Weatherley, N. and Jackson, A., 2005. *Cleaner Coasts Healthier Seas*. The State of the Marine Environment of England and Wales. Environment Agency, Bristol.

Barnett, R., 2005. Development opportunities in the mixed-use resort market. *Journal of Retail and Leisure Property*, 5(1). 17–23.

Booth, P., 1995. Zoning or discretionary action: certainty and responsiveness in implementing planning policy. *Journal of Planning Education and Research*, 14(2). 103–112.

Booth, P., 2002. Nationalising development rights: the feudal origins of the British planning system. *Environment and Planning B: Planning and Design*, 29(1). 129–139.

Bowen, A., Fankhauser, S., Stern, N. and Zenghelis, D., 2009. *An Outline of the Case for a 'Green' Stimulus*. Policy Brief, Grantham Research Institute on Climate Change and the Environment, London, February.

Boyle, M., McWilliams, C. and Rice, G., 2008. The spatialities of actually existing neoliberalism in Glasgow, 1977 to present. *Geografiska Annaler: B, Human Geography*, 90(4). 313–325.

Bradbury, J., 2005. Devolution: between governance and territorial politics. *Parliamentary Affairs*, 58(2). 287–302.

Bridgend County Borough Council, 2002. *Policy Statement on Flood and Coastal Defence*. Report of the Assistant Director of Environmental and Planning Services to the meeting of Cabinet, Bridgend, July.

Bridgend County Borough Council, 2004. *Porthcawl Development Framework; Supplementary Planning Guidance*. Bridgend County Borough Council, Bridgend.

Bridgend County Borough Council, 2007a. *Seven Bays Project. Porthcawl Waterfront. Supplementary Planning Guidance*. Bridgend County Borough Council, March.

Bridgend County Borough Council, 2007b. *Seven Bays Project. Report on Public Consultation Porthcawl Waterfront*. Bridgend County Borough Council, October.

Bridgend County Borough Council, 2007c. *Report to Cabinet for Information: Welsh Coastal Tourism Strategy*. Bridgend County Borough Council, 27 March.

Bridgend County Borough Council, undated. *Swansea Bay Shoreline Management Plan*. Bridgend County Borough Council.

Butler, R., 1980. The concept of a tourist area cycle of evolution: implications for management of resources. *Canadian Geographer*, 24. 5–12.

Carley, M., 2000. Urban partnerships, governance and the regeneration of Britain's cities. *International Planning Studies*, 5(3). 273–297.

Connell, L. and Flynn, A., 2003. The environment, innovation and industry: a case study of South Wales. *International Journal of Technology Management*, 17(5). 480–494.

David, J., 2006. *Porthcawl. Photographic Memories*. The Francis Frith Collection, Salisbury.

Davies, A. R., 2001. Hidden or hiding? Public perceptions of participation in the planning system. *Town Planning Review*, 72(2). 193–216.

Department for Environment Food and Rural Affairs, 2002. *Safeguarding Our Seas: A Strategy for the Conservation and Sustainable Development of our Marine Environment*. DEFRA, London.

English Heritage and Commission for Architecture and the Built Environment, 2003. *Shifting Sands: Design and the Changing Image of English Seaside Towns*. CABE, London.

Feeny, D., Hanna, S. and McEvoy, A. F., 1996. Questioning the assumptions of the 'Tragedy of the Commons' model of fisheries. *Land Economics*, 72(2). 187–205.

Fernie, K. and McCarthy, J. P., 2001. Partnership and community involvement: institutional morphing in Dundee. *Local Economy*, 16(4). 299–311.

Fudge, S., 2007. Objective 1 and community development: democratizing the economic and political landscape in South Wales? *Contemporary Wales*, 19(1). 1–15.

Fuller, C. and Geddes, M., 2008. Urban governance under neoliberalism: new labour and the restructuring of state-space. *Antipode*, 40(2). 252–282.

Gale, T., 2005. Modernism, post-modernism and the decline of British seaside resorts as long holiday destinations: a case study of Rhyl, North Wales. *Tourism Geographies*, 7(1). 86–112.

Gripaios, P., 2002. The failure of regeneration policy in Britain. *Regional Studies*, 36. 568–577.

Harris, N. and Hooper, A., 2004. Rediscovering the 'spatial' in public policy and planning: an examination of the spatial content of sectoral policy documents. *Planning Theory and Practice*, 5(2). 147–169.

Harris, N., Hooper, A. and Bishop, K. D., 2002. Constructing the practice of 'spatial planning': a national spatial planning framework for Wales. *Environment and Planning C: Government and Policy*, 20. 555–572.

Healey, P. and Shaw, T., 1993. Planners, plans and sustainable development. *Regional Studies*, 27(8). 769–776.

Higgins, L. S., 1968. *Newton Nottage and Porthcawl (From Prehistoric Times to 1950)*. Gomerian Press, Llandysul.

Holling, C. S., 2001. Understanding the complexity of economic, ecological, and social systems. *Ecosystems*, 4. 390–405.

House of Commons Communities and Local Government Committee, 2007. *Coastal Towns*. HC 351, Second Report of Session 2006–07, London.

Howe, J. and White, I., 2004. Like a fish out of water: the relationship between planning and flood risk management in the UK. *Planning Practice and Research*, 19(4). 415–425.

Institute for Public Policy Research, 2008. *Engagement and Political Space for Policies on Climate Change*. Institute for Public Policy Research, London, January.

Jeffery, C., 2007. The unfinished business of devolution: seven open questions. *Public Policy and Administration*, 22(1). 92–108.

Jennings, S., 2004. Coastal tourism and shoreline management. *Annals of Tourism Research*, 31(4). 899–922.

Kerr, G. and Johnson, S., 2005. A review of a brand management strategy for a small town – lessons learnt! *Place Branding*, 1(4). 373–387.

Kidd, S., Massey, D. and Davies, H., 2003. The ESDP and Integrated Coastal Zone Management: implications for the integrated management of the Irish Sea. *Town Planning Review*, 74(1). 97–121.

King, G., 2004. *Porthcawl Regeneration Area Sustainability Appraisal Final Report*. Environmental Planning Consultancy, London.

Lawless, P., 1994. Partnership in urban regeneration in the UK: the Sheffield Central Area Study. *Urban Studies*, 31(8). 1303–1324.

Lloyd, M. G. and Peel, D., 2008. Functionalism and representationalism in contemporary urban agendas: a constructive alignment? *Urban Research and Practice*, 1(1). 1–18.

MacLeod, G. and Goodwin, M., 1999. Space, scale and state strategy: rethinking urban and regional governance. *Progress in Human Geography*, 23(4). 503–527.

Marine Climate Change Impacts Partnership, 2008. *Annual Report Card 2007–2008*. J. M. Baxter, P. J. Buckley and C. J. Wallace (eds). Summary Report, MCCIP, Lowestoft.

McCarthy, J., 2007. *Partnership, Collaborative Planning and Urban Regeneration*. Ashgate, Aldershot.

McFadden, L., 2008. Exploring the challenges of integrated coastal zone management and reflecting on contributions to 'integration' from geographical thought. *The Geographical Journal*, 174(4). 299–314.

McGlashan, D. J., Duck, R. W. and Reid, C. T., 2004. The foreshore: geographical implications of the three legal systems in Great Britain. *Area*, 36. 338–347.

McGlashan, D. J., Duck, R. W. and Reid, C. T., 2005. Defining the foreshore: coastal geomorphology and British laws. *Estuarine, Coastal and Shelf Science*, 62. 183–192.

Morgan, A., 1975. *Porthcawl: Its History and Development*. D Brown and Sons Limited, Cowbridge and Bridgend.

Myatt, L. B., Scrimshaw, M. D. and Lester, J. N., 2003. Public perceptions and attitudes towards a current managed realignment scheme: Brancaster West Marsh, North Norfolk, UK. *Journal of Coastal Research*, 19(2). 278–286.

Newell, E. and Watts, S., 1996. The environmental impact of industrialisation in South Wales in the nineteenth century: 'Copper Smoke' and the Llanelli Copper Company. *Environment and History*, 2(3). 309–336.

Office of the Deputy Prime Minister, 2002. *The Learning Curve: Developing Skills and Knowledge for Neighbourhood Renewal*. Office of the Deputy Prime Minister, London.

Office of the Deputy Prime Minister, 2004. *Planning Policy Statement 1: Delivering Sustainable Development*. Office of the Deputy Prime Minister, London.

O'Riordan, T., Nicholson-Cole, S. A. and Milligan, J., 2008. Designing sustainable coastal futures. *Twenty-First Century Society*, 3(2). 145–157.

Parker, C. and Garnell, C., 2006. Regeneration and retail in Liverpool: a new approach. *Journal of Retail and Leisure Property*, 5(4). 292–304.

Parliamentary Office of Science and Technology, 2009. *Coastal Management*. POST, London, October, No. 342.

Peel, D. and Lloyd, M. G., 2004. The social re-construction of the marine environment: towards marine spatial planning. *Town Planning Review*, 75(3). 359–378.

Peel, D. and Lloyd, M. G., 2007. Civic formation and a new vocabulary for national planning. *International Planning Studies*, 12(4). 391–411.

Peel, D. and Lloyd, M. G., 2008. Governance and planning in the marine environment: regulating aquaculture in Scotland. *The Geographical Journal special issue on 'Coastal and Marine Governance'*, 174(4). 361–373.

Phillips, M. R. and Jones, A. L., 2006. Erosion and tourism infrastructure in the coastal zone: problems, consequences and management. *Tourism Management*, 27. 517–524.

Pitt, M., 2008. *Learning Lessons from the 2007 Floods*. Sir Michael Pitt Independent Review, Final Report, London.

Potts, J. S., 1999. The non-statutory approach to coastal defence in England and Wales: Coastal Defence Groups and Shoreline Management Plans. *Marine Policy*, 23(4–5). 479–500.

Powell, K., 2001. Devolution, planning guidance and the role of the planning system in Wales. *International Planning Studies*, 6(2). 215–222.

Poxon, J., 2000. Solving the development plan puzzle in Britain: learning lessons from history. *Planning Perspectives*, 15. 73–89.

Pugh, D. T., 2004. *Changing Sea Levels: Effects of Tides, Weather, and Climate*. Cambridge University Press, Cambridge.

Royal Town Planning Institute, 2009. *Planning to Live with Climate Change*. RTPI, London.

Royal Town Planning Institute, 2010. *Learning Outcomes for Accredited Course: Consultation*. RTPI, London.

Scottish Coastal Forum, 2004. *A Strategy for Scotland's Coast and Inshore Waters*. Scottish Coastal Forum, Edinburgh.

Scottish Environment LINK, 2006. *Marine Bulletin 1 – A Coastal and Marine National Park for Scotland*, Edinburgh.

Scottish Executive, 2005. *Seas the Opportunity: A Strategy for the Long Term Sustainability of Scotland's Coasts and Seas*. Scottish Environment LINK, Edinburgh.

Selby, M. and Morgan, N. J., 1996. Reconstructing place image. A case study of its role in destination market research. *Tourism Management*, 17(4). 287–94.

Slater, A.-M., 2007. A marine act: implications for Scotland. *The Journal of Water Law*, 18. 5–13.

Stern, N., 2007. *The Economics of Climate Change*. HM Treasury, London.

Turner, R.K., 2000. Integrating natural and socio-economic science in coastal management. *Journal of Marine Systems*, 25(3). 447–460.

Walker, D., 2008. Sustainability: environmental management, transparency and competitive advantage. *Journal of Retail and Leisure Property*, 7(2). 119–130.

Welsh Assembly Government, 2004. *People, Places, Futures. The Wales Spatial Plan*. Welsh Assembly Government, Cardiff.

Welsh Assembly Government, 2007. *Welsh Coastal Tourism Strategy. Draft Final Strategy Document*. Welsh Assembly Government. Visit Wales, the Tourism and Marketing Division, Cardiff.

Welsh Assembly Government, 2008. *People, Places, Futures. The Wales Spatial Plan 2008 Update*. Welsh Assembly Government, Cardiff.

Whitehead, M., 2007. The architecture of partnerships: urban communities in the shadow of hierarchy. *Policy and Politics*, 35(1). 3–23.

For Product Safety Concerns and Information please contact our EU representative GPSR@taylorandfrancis.com Taylor & Francis Verlag GmbH, Kaufingerstraße 24, 80331 München, Germany

Batch number: 10733608

Printed by Printforce, the Netherlands